TABLE OF CONTENTS

\\Foreword _9

\\Introduction _11
\\Tendering requirements _12
\\Costs _13
\\Deadlines _15
\\Function _16
\\Range of services _18
\\Quality _19
\\Tender items _20

\\Organizing the tender _23
\\Timetabling the invitation to tender _23
\\Time invested by participants _24
\\Tendering sequence and nature of tender _26
\\Fixing bid units _28
\\Tendering by trade _28
\\Invitation to tender by part lot _31
\\Package awards _31
\\Tendering style _33
\\Tendering by function _33
\\Detailed tendering _34
\\Depth of tender _34

\\Structuring an invitation to tender _37
\\Textual elements _37
\\General information about the building project _37
\\Contractual conditions _41
\\Technical requirements _44
\\Project-related contractual conditions _46
\\Tender specification _47
\\Drawing elements _48
\\Other descriptive elements _49

\\Structuring service specifications _51
\\Approach to functional service specifications _51
\\Functional tender specifications without a design _53
\\Functional specification with design _57
\\Structuring a functional tender specification _60
\\Drawing up a requirement profile _62

\\Procedures for a detailed description of works _66
\\Lot _69
\\Title and subtitle _69
\\Items _71

\\In conclusion _82

\\Appendix _83
\\Literature _83
\\Sample International Contracts _83
\\Additional sources of information _83
\\Picture credits _84

前言

建筑的质量不仅来自于好的、创造性的设计，而且和建造的过程密切相关。因此对于建筑来说，从第一张设计草图到最后的结构和表面设计，质量都是一个贯穿始终的问题。招标是设计和施工之间的桥梁，界定了项目的要求。相关的承包人通过标书条款熟悉项目的细节，然后提交一份描述工作进展的标书，之后这份标书就会成为合同和实际建筑任务书的基础。从这个角度来说，作为建设开始前最后一个设计阶段的标书条款，是一个非常重要的设计组成部分。所以它的概念必须具有像设计或工作计划一样的精确性和可靠性。

由于学生和专业设计师的职业经验比较少，所以他们需要经过认真组织的、实用的、关于招投标类型、过程和单项服务的信息来帮助他们完成第一份标书。本书从这个非常初级的阶段开始，简单易懂地介绍和解释了整个过程。

首先，它描述了如何在标书中详细说明工作的内容和解释过程中的基本原则。它讨论了分包的类型、确定中标单位、时间安排以及非常重要的标书条款的风格。正如本书所展示的那样，对工作的说明不仅要有单纯的功能性描述，还要有非常详细的内容。读者可以看到招标过程中的各个组成部分，它们的意义何在，以及如何进行组织。在完成标书的过程中，这是要以很多的实践经验、实例以及简单而清晰的总结为支撑的。本书提供了开始一份合理而实际的投标工作之前所需要的所有关键背景和内容。

编者：贝尔特·比勒费尔德

FOREWORD

Architectural quality does not derive from good and creative designs alone, it must also be reflected in the built reality. So in architecture quality is a continual concern, from the first sketch via design and final planning to built structures and surfaces. Tendering is the link between planning and realization, defining the project requirements. The responsible contractors use the tender specifications to familiarize themselves with the details of the commission and can then submit a tender to carry out the work described, which then forms the basis for the contract and the actual building programme. In this way the tender specification, as the final design development stage before building starts, is an important planning component. It should therefore be compiled with the same precision and faithfulness to the concept as a design or working plan.

As students and professional beginners usually have little professional experience, they need carefully structured, practical information about tender types, the sequence of events, and descriptions of the individual services required to help them handle their first tender specifications. The present volume starts at this early stage and works through the subject matter with the aid of readily understood introductions and explanations.

First of all it describes how building services are specified in a tender, and explains the fundamental principles of the process. One important practical topic is organizing tendering for a building project. It discusses types of award, fixing award units, time planning and not least the style of the tender specification. The possibilities for describing the work required range from the purely functional to more detailed description, as will be shown. Readers discover the components of the tendering process, what they are there for and how they should be compiled in detail. This is supported by practical tips, examples and simple, clear summaries that help when drawing up a tender specification. The *Tendering* volume of the *Basics* series passes on all the key background and context needed for a sound and practical start on tendering work.

Bert Bielefeld
Editor

ARCHITECTURAL COMPOSITION
WELT
ARCHITEKTUR.de
767,00
6.085,96
728,00
427,00
1.394,75
3.150,00
206,00
12.751,71
12.751,71
2.041,32
14.800,10
Lippstadt, 25.08.06
3.363,25
13,10
32,50
426,13
29,50
536,90

INTRODUCTION

From planning to realization

Planners – this includes architects, civil engineers and specialist engineers – have to involve another group of people by the end of the plan submission and planning permission process at the latest. The earlier phases of the process focused on communicating with the client and the authorities, but now planners have to turn their attention towards the firms who will be responsible for realizing the project: craftsmen and women, building contractors, specialist companies.

Tender content

All the information needed for realizing the project is provided to building firms as part of an invitation to tender, in the form of descriptions or drawings of the work to be done and services needed. The invitation to tender must contain all the information the bidding firms need to perform the necessary services and to submit a bid for the contract and, where appropriate, for planning the work.

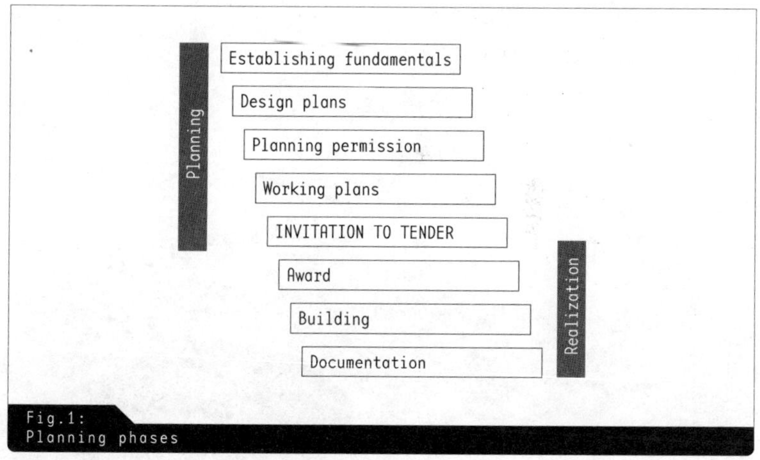

Fig.1: Planning phases

\\Hint:
The tender becomes part of the contract that the client concludes with the building contractor. Planners can incur penalties for errors and omissions in the bid.

Outcome of the tendering process

The tendering process aims to attract as many appropriate bids as needed to form a broad view of the market. The invitation to tender is compiled by the planner and submitted to suitable contractors, who then calculate prices and submit a bid, which is binding. This is then examined by the planner and compared with other bids. The comparison gives the planner an insight into current prices and enables the client to commission the work from the bidder who has submitted the most reasonable offer for the particular project.

TENDERING REQUIREMENTS

The invitation to tender collects all the requirements that have come to light in the planning phase. These requirements are essentially laid down by the client, but they may also relate to legal or technical matters. They can be categorized as follows: › Fig. 2

- Costs
- Deadlines
- Function
- Scope
- Quality

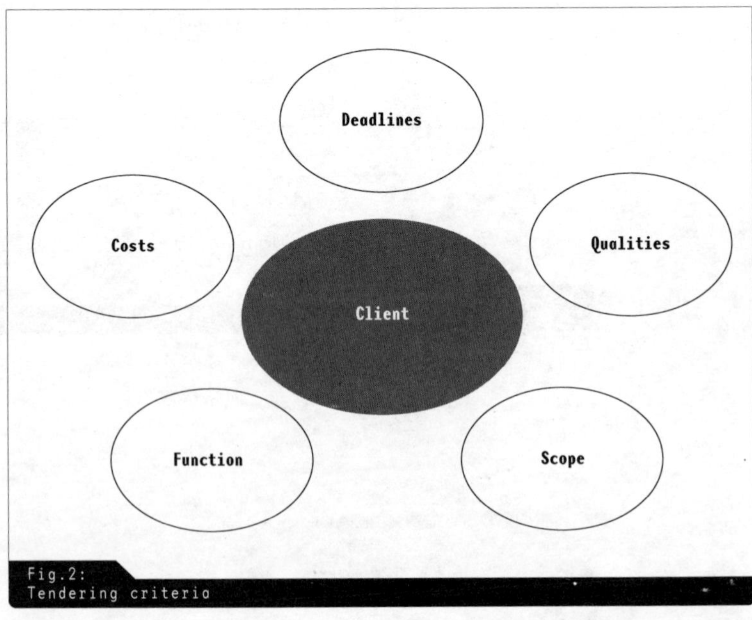

Fig.2:
Tendering criteria

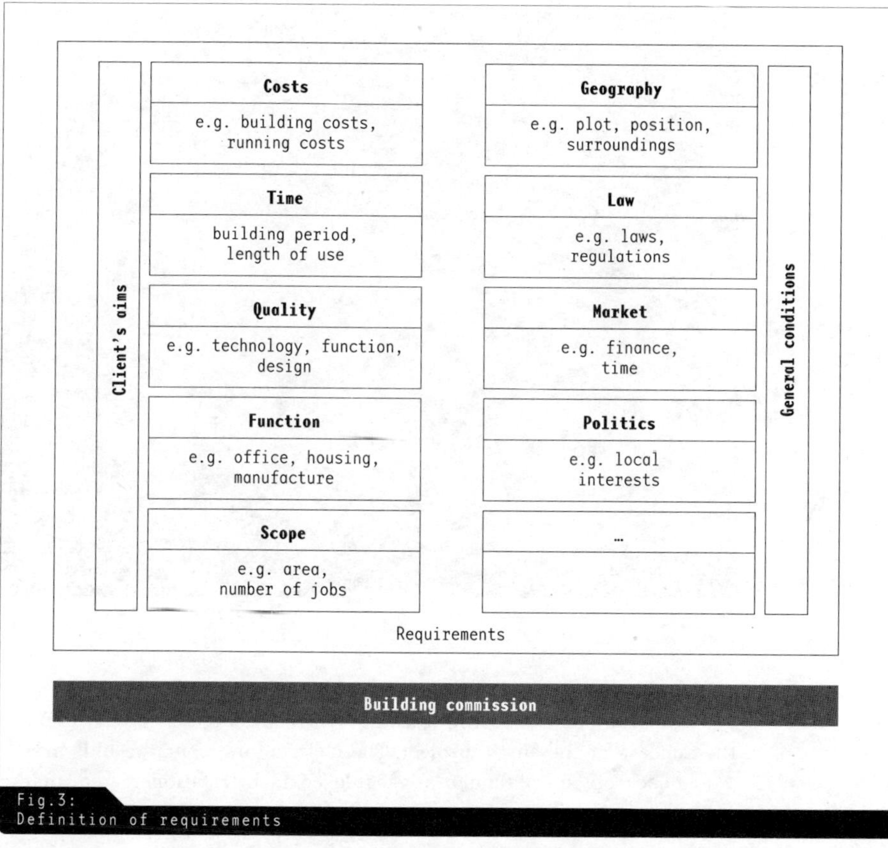

Fig.3: Definition of requirements

These criteria are used to fix the realization phase and to identify any planning services that are still required. › Fig. 3

Costs

Cost range

In most cases cost is the key criterion for or against a realization bid, or even for or against the building project itself. Planners are obliged to spend a client's money on the project in the client's best interest. Planners generally have a prescribed budget, and must set all the costs arising from the building work against this. This will mean drawing up separate budgets for the various service packages or award units. › Chapter Organizing the tender, Fixing bid units

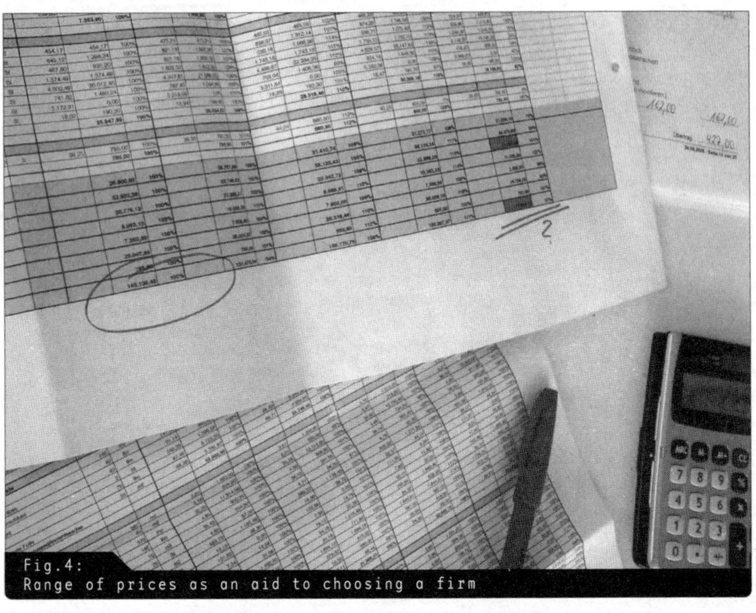

Fig.4:
Range of prices as an aid to choosing a firm

For the client, keeping to the prescribed budget is often crucial to the success of the entire project. The contractors' concrete bids are the first check on the planner's suggested costs in relation to real market prices.

\\Tip:
Bids submitted in response to the first invitation to tender are crucially important in establishing the client's confidence in the planner's costing competence. If even the first bids come in outside the planned cost framework, it is possible that the client, anxious about keeping within the overall budge, might make radical adjustments at an early stage that could affect all the other criteria, e.g. a marked reduction in the standards for the finished building.

Cost control

An overview of the individual budgets makes it possible for planners to control costs. If the bid for a particular unit is above the budget allocated to it, planners will have to cut the budget for other units and take this into consideration when drawing up invitations to tender, for example by reducing quality standards or the scope of the work required. Conversely, if an item comes in under budget, planners can, for example, include clients' requirements that had previously fallen outside the cost framework.

Cost guarantees

Clients can best ensure that costs are firmly fixed by attempting to eliminate all cost risks arising from unpredictable events during building, from market developments, and from submitting a series of individual tenders. One possible way of doing this is for a single contractor to take on the whole operation, which guarantees completion costs and deadlines.
› Chapter Organizing the tender, Fixing bid units, Package awards

Deadlines

Clients will generally set firm deadlines, or at least express their wishes about them. Once deadlines are agreed, they are binding.

Deadline constraints

Constraints on deadlines arise mainly from the planned use of the building concerned. For example, completion dates, and thus possible moving-in dates, are crucially important for private clients building their own home, who need to give appropriate notice on their previous, rented accommodation. Renovation work in schools can often be carried out only in the school holidays. Here both the starting and completion dates are deciding factors.

\\Tip:
Planners must establish that the client's ideas about deadlines are realistic. This affects the services that can be delivered by the firms involved, and the planners' own ability to deliver. Some events that occur in the course of building can be influenced only slightly or not at all. These include gaining permission from authorities, the weather, and some product delivery times.

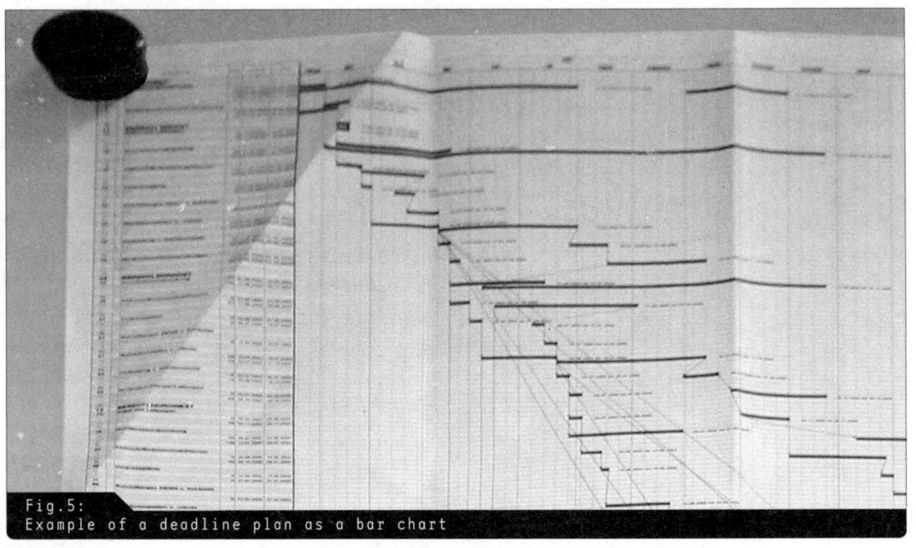
Fig.5:
Example of a deadline plan as a bar chart

Deadlines as a cost factor

Deadline requirements also influence possible construction processes and thus costs. The only realistic way of working faster is to employ a larger workforce, more machines and materials. Contractors could then be compelled to hire equipment or to complete the work in overtime, working at weekends or even at night. This will result in high bid prices, as the company factors the extra costs into the bid price.

Effects on tendering

Deadline requirements also affect the way planners submit their tenders. Robust and detailed planning involves investing a great deal of time, so planners have to consider whether they will be able to submit such plans at the appropriate time. If they cannot do so, they can transfer some of the planning services to the contractor, by defining some aspects in terms of functions, rather than in full detail. › Chapter Organizing the tender, Tendering style, Tendering by function

Function

Realization range

The client's requirements establish the extent and bandwidth of the realization variants. For example, if a private client wants to buy land and build a home on it, this can be a terraced house, a semi-detached house or a detached house. Function is thus one of the factors determining the form the building will take. It is also possible to decide on particular building

Fig.6:
Various functions

methods from a function description. When building a warehouse with no special requirements, a choice will usually be made between favourably priced variants (e.g. reinforced concrete or steel construction). Thus function is linked with a particular range of possible solutions, modified by the client's individual requirements.

Client profile

The more strongly clients identify themselves with a commission, the more influence they will wish to exert on planning the invitation to tender. If the project is their dream house, the client could well wish to be involved in every last detail of the planning process. The invitation to tender will thus have to be correspondingly detailed, so that the client's ideas can be implemented in full. › **Chapter Organizing the tender, Tendering style, Detailed tendering**

If the finished building is intended as a for-profit project, however, clients will be mainly interested in minimum costs for maximum yield. They will want to scale their requirements down as much as possible at first, and will be prepared to raise their technical or aesthetic sights only if there is a prospect of higher profits or greater marketability. If clients are simply after a box to put something in (e.g. a warehouse or an industrial production hall), they will also tend to see the commission pragmatically in terms of function, and not want to bother themselves with too much detail.

Fig.7:
There are various ways of meeting requirements for a building.

Range of services

Minimum scope

The range of services derives from clients' wishes. For example, when building an office block, clients can state how many office workstations are intended and what other spaces are needed to serve the desired function (foyer, conference rooms, server areas etc.). The more precise the requirements, the more precisely the minimum project range can be determined.

Rationalization

If the project range is inappropriate to the desired cost framework, planners can reduce costs by rationalization (for example, by using a large number of identical elements and focusing on the same service provider as much as possible): façade design can match the façade panel format the manufacturer produces, so that large quantities of a particular panel format can be used without cutting or having to order special formats.

Factors open to influence

As well as the scope set by minimum standards, there are also variable quantities that affect the quality of the building as a rule. For example, planners can minimize the window area, which is more expensive than a closed façade, at the expense of user comfort, or reduce the number of workstations at the expense of subsequent flexibility.

Fig. 8:
The relationship between function and quality

Quality

Function also affects the quality expected. Here we can speak of technical and aesthetic criteria. › Fig. 9 Technical requirements include building law provisions (e.g. statutes relating to assembly of persons or the fire prevention concept), or health aspects (e.g. ventilation or hygiene); aesthetic requirements relate to the visual impact, form and characteristics of the building as a whole, down to individual details such as door handles.

Standards of quality

There are fixed minimum standards for most building services, intended to guarantee the use of appropriate materials and professional execution. Clients will have requirements for their property that go beyond minimum quality. As soon as the planned finish deviates from the standard quality, planners must mention this expressly in their service description and describe the finish or the desired result.

If requirements relating to quality of finish exceed the normal standard, costs will rise as well. For example, the amount of work and cost involved in dry-building a wall with a high level of overall finish on the plasterwork is considerably greater than for one that is simply smoothed and finished at the joints.

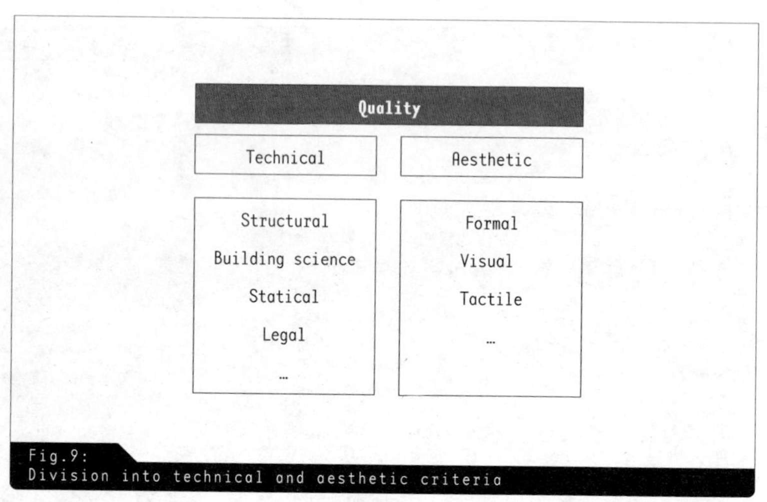

Fig.9:
Division into technical and aesthetic criteria

Longer-term planning

Long-term considerations should always be included in relation to quality. For example, installing a more expensive but higher-calibre heating system can easily compensate for its greater cost through lower energy costs over the period of use.

TENDER ITEMS

Building services and construction products

The building process involves choosing and coordinating an enormous range of structural elements. Here, planners have a very wide range of prefabricated items at their disposal (e.g. doors and door frames), but can also work with individually manufactured elements (e.g. hand-crafted door fittings). A building can be planned down to the position of the last screw, and the shape of its head. Invitations to tender for building services relate both to parts of the planning process and also to the whole realization process. They summarize all the services needed. The scope of the invitation to tender will vary according to the scope of the building project and the nature of the tender. › Chapter Organizing the tender, Tendering style For example, if tenders are being invited for a complete building project, starting from scratch to the very end, they can include all construction services from digging the foundations to cleaning at the end of the construction phase and handing the key over to the client. Invitations to tender may also be issued for replacing a single window.

Fig.10:
Everything can be built into an invitation to tender

Planning and other services

Complementary planning and services can also be included in an invitation to tender, as well as classical construction work or products. So it is possible to invite bids for specialist planning such as preparing a sound insulation report, or services such as organizing a topping-out party.

Das Bauhaus
TENNISCLUB >GRÜN-WEISS< 1911 E.V. LIPPSTADT

ORGANIZING THE TENDER

Fundamentals of organization

The possible scope of the project, and the diversity and complexity of the invitation to tender, mean that it makes sense to divide the building process into significant phases. To do this, planners must be familiar with events within the construction process and the individual events' interdependence, so that they can arrange them in the correct time sequence.

TIMETABLING THE INVITATION TO TENDER

Deadlines and tendering

Deadlines are an important element of the tendering process. Planners have to know what periods of time are realistic for realizing the project. A timetable for the planner's and the bidding companies' work on the tender can be drawn up with reference to the realization deadlines,

\\Hint:
The process following the invitation to tender, collecting in bids and subsequently commissioning of firms by clients, is called the awarding procedure.
Information about awarding tenders can be found in *Building projects in the European Union* by Bert Bielefeld and Falk Würfele, Birkhäuser Verlag, 2005.

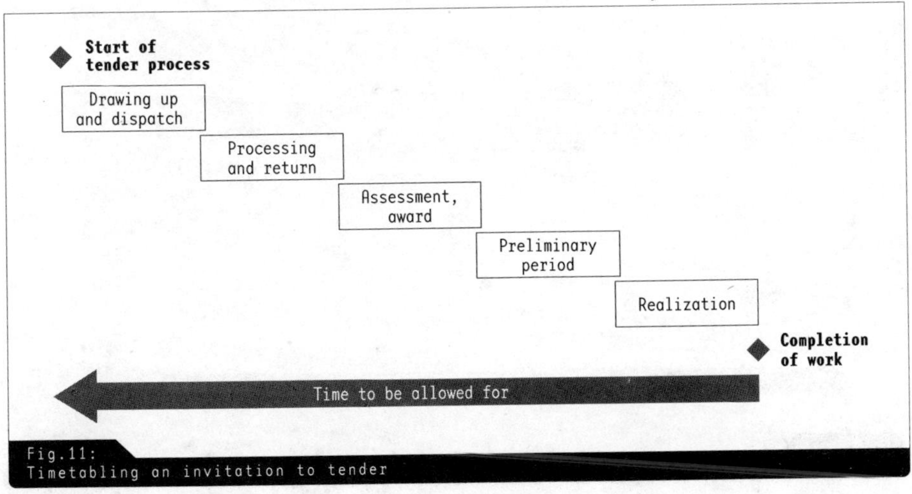

Fig.11: Timetabling an invitation to tender

bearing possible preliminary planning periods for specialist firms and for the awarding procedure in mind.

Time invested by participants

Time invested by planner

Planners need sufficient time to draw up an invitation to tender. ›Fig.11 Once they have compiled a list of all the client's wishes and requirements, they must take time to organize the invitation to tender and think how to convey the requirements in such a way that the invitation can be formulated meaningfully. Planners must establish quantities needed, to define the scope of the services required. Any question arising must be cleared up with manufacturers, specialist organizations or other appropriate contacts. It is often necessary to provide any experts approached with documents about the general conditions, and drawings, to ensure that responses are robust and appropriate for describing the services required. If difficult installations or complex construction processes are involved it often makes sense for planners to cover themselves by asking manufacturers for written statements or opinions.

Once planners have drawn up their lists of services needed, they must compile a list of bidders, i.e. the names of all the firms invited to tender, in consultation with the client where appropriate. The tender documents

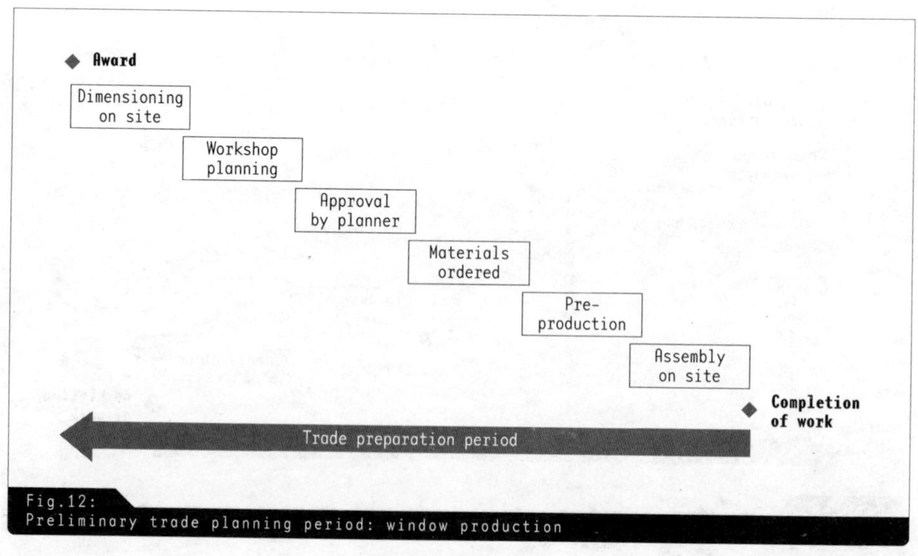

Fig.12:
Preliminary trade planning period: window production

have to be duplicated and sent out to the companies concerned, allowing an appropriate length of time for processing by the companies.

Time needed for processing the bid

Construction companies submitting bids who receive a description of the services required have to familiarize themselves with a new building project, and thus need an appropriate period to work on the invitation to tender. In some cases, the nature and scope of the tender may require additional planning work before the price can finally be calculated; manufacturers or other firms and their internal price enquires may need to be considered in their turn. The calculation must take wages, materials, equipment and outside services into account. As well as these factors relating directly to the building commission, general overheads and possible profit have to be built into the bid price. Preparing the bid can take anything from a day to several weeks, according to the complexity and scope of the tender. The necessary processing period is extended correspondingly if the invitation to tender also covers planning services or technical tests.

Preliminary planning period for trades

The time span from commissioning to the actual delivery of the services (start of building work) on the building site is the preliminary planning period for specialist firms. ›Fig. 12 During this period, the firms being commissioned can construct working plans or samples and submit them to the planner for approval. Construction elements are often modified or

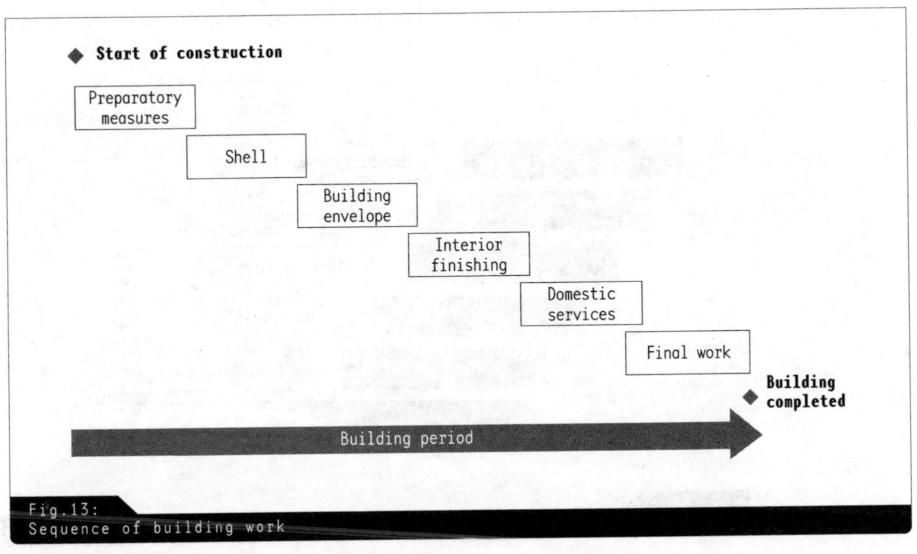

Fig.13:
Sequence of building work

assembled in advance in the factory by the firms involved. Measurements for prefabrication may need to be made on site, which adds another factor for meeting completion deadlines, for example if masonry with apertures has to be completed before fixing the dimensions of the windows that have to be prefabricated.

Time needed for realization

When timetabling tenders it is important to know how long it will take to complete a particular piece of work, given that a possible completion deadline has to be fixed. ›Fig. 13 There is only limited scope for shortening such an individual completion time. The length of time needed can be affected by the number of people working on the job, working hours and the use of machines. There are natural restrictions on speeding up work, for example the time that certain building materials take to dry or harden (e.g. screed). Space on the building site may be at a premium, so increasing the workforce could mean people getting in each other's way while working.

TENDERING SEQUENCE AND NATURE OF TENDER

Time sequence

The order for drawing up invitations to tender is usually based on the order in which work is carried out on the building site. ›Fig. 13 First bids are invited for preparatory measures for the actual building project, followed by bids for shell realization, exterior finish, interior finish and fittings, down to bids for the final work needed. Sometimes the preliminary preparation period makes it necessary to deviate from the on-site sequence when inviting to tender. For example, façade construction can

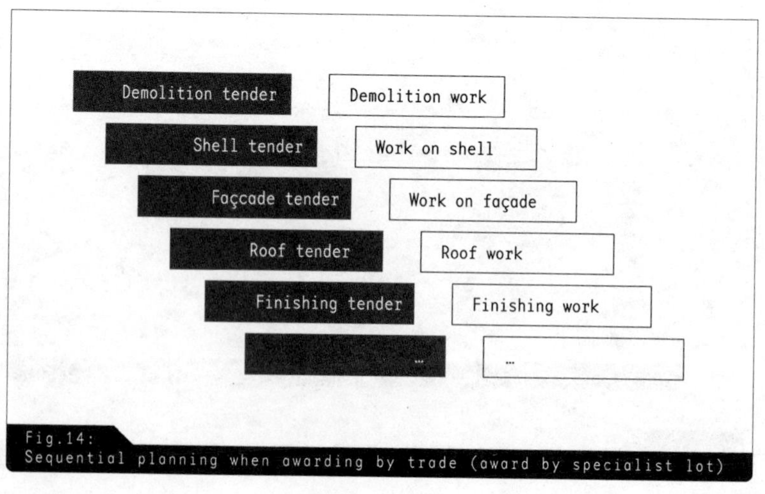

Fig.14:
Sequential planning when awarding by trade (award by specialist lot)

entail a preparatory period lasting several months, which must be taken into consideration when drawing up the invitation to tender.

Invitations to tender while building is in progress

The sequence described here relates to invitations to tender issued while building is in progress. ›Fig. 14 For tendering in this way, all the services required are drawn up in sequence, and invitations to tender issued. For example, the interior is planned and invitations to tender are issued after the shell has already been completed. In comparison with a blanket invitation to tender ›Chapter Organizing the tender, Fixing bid units, Package awards this approach offers the advantage that an appropriate response can be made to unexpected cost developments. ›Introduction, Tendering requirements, Costs It is also possible to accommodate changes that have occurred during the completed building phases. For example, if the ceiling slab thicknesses have had to be changed for practical reasons, they can be compensated for in the finished floor height. However, it is impossible to be certain about costs until the last invitation to tender, because of the difficulty in predicting the effect market fluctuations and other eventualities could have on services offered.

Blanket tendering

Awarding to a main contractor offers greater cost security. ›Chapter Organizing the tender, Fixing bid units, Package awards Here, all the services have to be identified in full, and submitted to the contractor with an invitation to tender. The planning period involved is correspondingly long. Planning previously undertaken in parallel with the building work now has to be completed before the first ground is dug. ›Fig. 15 If there is not enough time available for a detailed invitation to tender, planners must concentrate

Contractor with detailed invitation to tender

Planner	Contractor
Planning	Realization

Contractor with functional invitation to tender

Planner	Contractor	
Planning	Planning	Realization

Fig.15:
Award procedure and time consumed

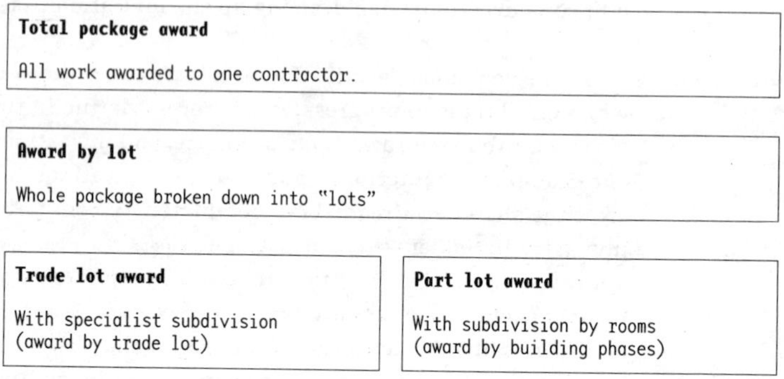

Total package award

All work awarded to one contractor.

Award by lot

Whole package broken down into "lots"

Trade lot award

With specialist subdivision (award by trade lot)

Part lot award

With subdivision by rooms (award by building phases)

Fig.16:
Definition of the terms for award units

on requirements that are important to the client and describe only these in detail › Chapter Organizing the tender, Tender style, Detailed tendering and the rest merely functionally. › Chapter Organizing the tender, Tendering style, Tendering by function Sometimes the pressure of time can be so great that a purely functional invitation model has to be considered, giving no detail at all.

FIXING BID UNITS

Bid unit

A bid unit defines the range of service provision awarded to a particular contractor. Bid units can be itemized according to size for individual trades (trade lots), part lots and complete packages; it is also possible to include all the services in a total package award. › Fig. 16

Tendering by trade

Subdivision by trades (trade or specialist lots) is based on craft and technical skills traditionally delivered by an individual or a firm (e.g. craft trades such as stonemason, carpenter or screed layer). This is generally the smallest bid unit. › Fig. 17

Trade (specialist lots)

A trade can be broken down into even smaller units if a number of different services are provided. For example, a tender invitation for metalworkers could include all the services this trade offers. It is also possible to draw up several invitations to tender identifying individual items, such as a service described under façade construction in metal, and another for

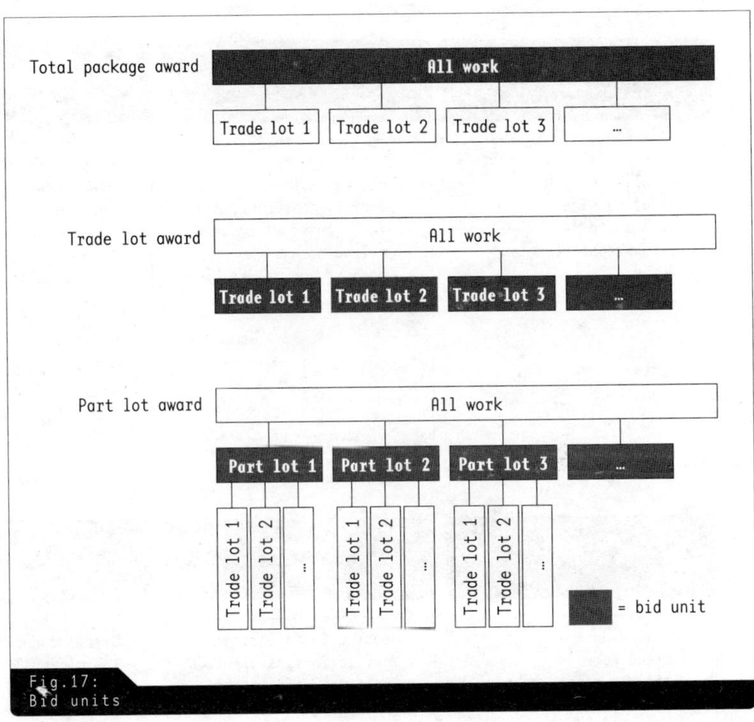

Fig.17:
Bid units

steel staircases and banisters. The smallest possible bid unit is a single service. › Fig. 18

So tendering by trade may include all the services performed by that trade or just some of them. It makes sense to break a trade down into smaller units if particular firms' specialist fields are to be used.

\\Hint:
The term trade is also generally applied to less traditional work such as structural engineering, media planning or sign-making. Although these are not traditional trades, the important feature here is that the services form a unit.

\\Example:
A construction company that produces and assembles stairs every day can offer this service more professionally and possibly more cheaply than a metalworker who specializes in façades but theoretically covers all aspects of that trade.

Preparatory measures	Shell	Building envelope
- Building site preparation - Demolition work - Clearing site - Excavations - Site preparation - …	- Excavations - Masonry work - Concrete construction work - Steel construction work - Sealing work - Carpentry and timber work - Scaffolding work - …	- Carpentry and timber work - Steel construction work - Sealing work - Roofing work - Plumbing work - Heat insulation work - Plastering - Façade work - Metal construction work - Glazing work - Painting - Scaffolding work - …

Finishing	Domestic services	Final measures
- Plastering - Screed work - Floor covering work - Concrete block work - Natural stone work - Tiling and - Parquet laying - Metalwork - Dry construction work - Joinery - Painting - Scaffolding work - …	- Heating installation - Ventilation installation - Sanitary installation - Electrical insulation - Lifts - Media technology - …	- Cleaning building - Installing locks - Outside areas - Clearing site - …

Fig.18:
Typical trade subdivisions

The disadvantage is that more time and effort have to be invested in coordination when commissioning several firms, and synergies (e.g. travel to the building site or larger delivery quantities at correspondingly more favourable prices) could be lost.

Bundling trades

It can sometimes make sense to bundle a number of trades. It seems logical to commission a single firm to take on all the work relating to a roof, and avoid having to coordinate a number of firms. Thus, carpentry (constructing the roof truss), roof-covering work (roof construction from insulation to the pantiles), and some metal-fitting work (fitting gutters, protective leading) can all be done by the same firm. Many firms have adapted to the clients' desire to deal with a single contact person, and advertise as providing a complete service. Note here that some firms that seem quite large simply "buy in" services and often cannot offer them at particularly reasonable prices. The client is then buying convenience by paying an additional price for in-house subcontractor organization by the firm commissioned. › Chapter Organizing the tender, Fixing bid units, Invitation to tender by part lot

Invitation to tender by part lot

Part lot

The part lot is another bid unit. Here, services are not classified in terms of a specific trade, but structured in sections. These sections derive mainly from a desire to be able to award to several firms when a great deal of work is involved.

Subdividing services

In public commissions, this can take place with the intention of involving as many firms as possible in the bidding process, as the scope of services required will then be based on the capacities of essentially average companies.

Building phases

Another sensible reason for structuring in part lots is when planning work over a long period with possible interruptions. Building phases are often fixed for larger building projects so that some parts of the building can be used while others are completed at a later stage. › Fig. 19

Package awards

Main contractor tendering

Bundling several trades, with only one contact person on the realization side, as mentioned above, is pursued further in awards to a main

\\Hint:
A subcontractor works for the firm with which the client has signed the building contract. The subcontractor has no official relationship with the client. The commissioned firm remains responsible for commissioning, finish, payment and guarantees.

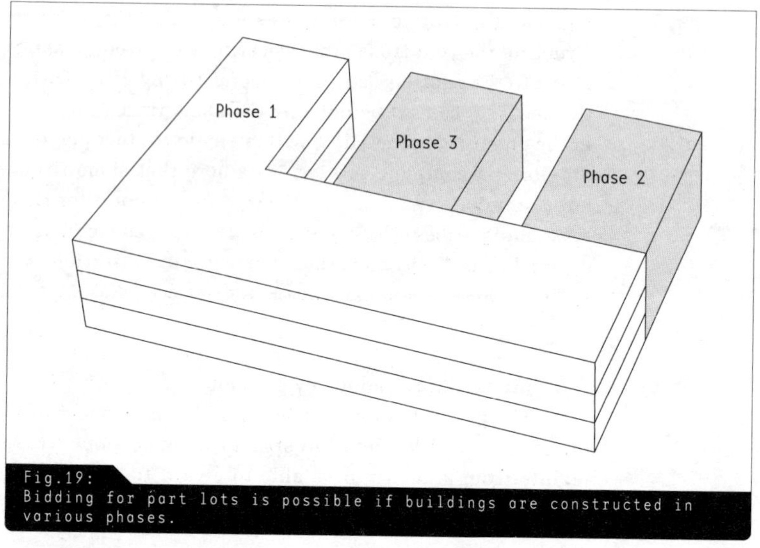

Fig.19:
Bidding for part lots is possible if buildings are constructed in various phases.

contractor. › Chapter Organizing the tender, Fixing bid units, Invitation to tender by part lot Here, the client commissions a single building firm to provide all the services needed to complete the building work. A single contract is agreed, rather than a large number of them.

Meeting deadlines

It is easier for a main contractor to fix completion dates because such a firm will be able to compensate for delays in parts of the project by pushing the work ahead in other areas, as part of the overall coordination process. It is more difficult to set binding deadlines when commissioning a number of individual firms because of the large number of mutual dependencies: individual building firms are not contractually obliged to each other.

Main contractor supplement

A main contractor is responsible for all this, and will generally exact payment for this often voluminous coordination work, and sometimes also risk coverage for guarantees undertaken, by building supplements into the bid. In practice there are few firms that can cover all the services required using in-house workers. In fact, they tend to tender the services out to other firms, which then – if they are commissioned – work as subcontractors. A main contractor's bid concludes in a guaranteed price for which the services must be delivered by a contractually fixed deadline.

TENDERING STYLE

A distinction is made between functional and detailed tendering, but these are rarely separated consistently in practice. Any detailed invitation to tender will always contain functional elements. For example, even a meticulously detailed description of a plasterboard stud wall will not contain precise information about fixing the plasterboard panels. It is assumed that the workmen will have the appropriate technical knowledge and will know the correct screws to use for fixing the panels to the frame. A functional invitation to tender can work without detailed elements, but here, too, there will in practice be areas where the requirements are formulated in greater detail. The more questions the planner asks the client about requirements, the longer the list of detailed requirements within the actual functional invitation will become.

Tendering by function

Functional tendering does not describe how the work is to be done or the precise building process, but focuses on the required outcome. The bidder takes responsibility for planning the work and thus also carries the risk of achieving the required result even if there were omissions in the original bid. As well as being responsible for possible planning errors, the contracted firm also carries the quantity surveying risks.

Bidders are able to determine how the work is done by choosing procedures in the light of their expertise and experience. They can optimize the entire range of services offered in terms of their own resources, as the contract offers room for manoeuvre.

Assessing bids

The criteria for assessing bids include price, and the way the set requirements are addressed. The bidder will have spent time and effort on the bid, and the planner now has to assess it in some depth. Consequently, clients or planners have no further influence in principle on the subsequent execution of the process. This loss of control, which applies to detailed planning in particular, may lead to a loss of design quality.

Choosing functional tendering

Functional tendering is often chosen through lack of time. ›Introduction, Tendering requirements, Deadlines It thus clearly reduces the extensive planning process that would have to precede award to a main contractor. Lower client demands on the realization details may lie behind a functional invitation to tender, especially as the firm to which the contract is awarded takes on a large number of risks as well. Another reason for choosing a functional style may be simply that the planner has no idea how to achieve the required aims by means of a detailed invitation to tender. Thus, planners will not invite tenders for the individual components of an air-conditioning plant or the way they are assembled, but will simply describe cooling or ventilation rate requirements.

Detailed tendering

Detailed tendering requires every detail of the work required to have been planned in advance to the greatest possible extent. Planners do not simply describe the required result, but also how it is to be achieved. They thus accept the risk that the finished work will not meet demands, or that there will be errors and omissions in the tender invitation, or it will not be completely clear. This can lead to additional costs for additional work (services that are needed but were not included in the original invitation to tender).

Assessing the bids

It is much simpler to assess a detailed invitation to tender, as the choice of procedure is fixed, and only the prices have to be compared.

Choosing a detailed invitation to tender

It always makes sense to opt for a detailed invitation to tender if the client wishes to remain in control of the building process. This is the only way of checking every detail of the realization work, and avoids disagreeable surprises.

Depth of tender

It is fundamentally possible to mix functional and detailed tendering. This opens up considerable creative possibilities for planners. They will be able to submit detailed final working plans for all the areas that are important to clients, and to describe the realization process with equal precision. In areas that do not require so much detail they can confine themselves to describing requirements and choose the contractor who offers the best possible solution.

Detailed or functional?

If planners put out detailed invitations to tender they must have the appropriate knowledge at their fingertips. They will be responsible for any mistakes in their description of the services they are offering. It is therefore advisable to tender on the basis of function, bearing the desired result in mind, for any elements about which they are not thoroughly informed.

Completeness of the bid

Planners must always ask themselves whether the invitation to tender they have prepared is complete, in other words whether the information they have provided is unambiguous, and that there are no omissions.

For example, if they ask for an "orderly and symmetrical" pattern of screws for securing the façade elements, they must add a diagram showing the pattern of screws, to avoid contentious interpretations of this requirement. As a rule, only detailed descriptions allow control of the way the work is ultimately done. This is very time-consuming, and cannot always be managed for every aspect of the building. Planners should always consider carefully what degree of detailing is necessary and appropriate. For example, requirements about formwork for an exposed concrete wall must be much more carefully formulated than those relating to formwork for foundations that will not be visible when the building is completed.

P 212
Dachdeckungs- und Klempnerarbeiten
Grundlagen
Mengenermittlung
P 212
Dachdeckungs- und Klempnerarbeiten
Ausschreibung
Anschreiben
HLSE
P 212
Schlosserarbeiten II
Auswertung
P 212
Tischlerarbeiten
Grundlagen

STRUCTURING AN INVITATION TO TENDER

The invitation to tender – functional or detailed – is made up of several elements. › Figs. 20, 21 It includes all the documents required for awarding a building contract.

TEXTUAL ELEMENTS

Textual elements mean all the descriptions couched in words and figures that provide information about the sequence and execution of the planned building project. An invitation to tender is drawn up using these elements, and consists of the following components:

_ General information about the project
_ Contractual conditions
_ Technical requirements
_ Information about building site conditions
_ Description of the work required

Textual elements usually make up a large proportion of a tender bid. Words enable planners to convey information that cannot be found in the plans.

General information about the building project

General description

A complete tender package contains general information about the planned building project and the awarding procedures. This information is conveyed in a cover sheet or accompanying letter containing the invitation to tender and other details relevant to the award, as well as the names of the key participants and a short description of the building.

\\Hint:
For an invitation to tender to be absolutely unambiguous it is important for individual elements not to contradict each other; this may be avoided by fixing a rank order for the individual elements. Thus, a precise description of the services required always ranks higher than the technical requirements.

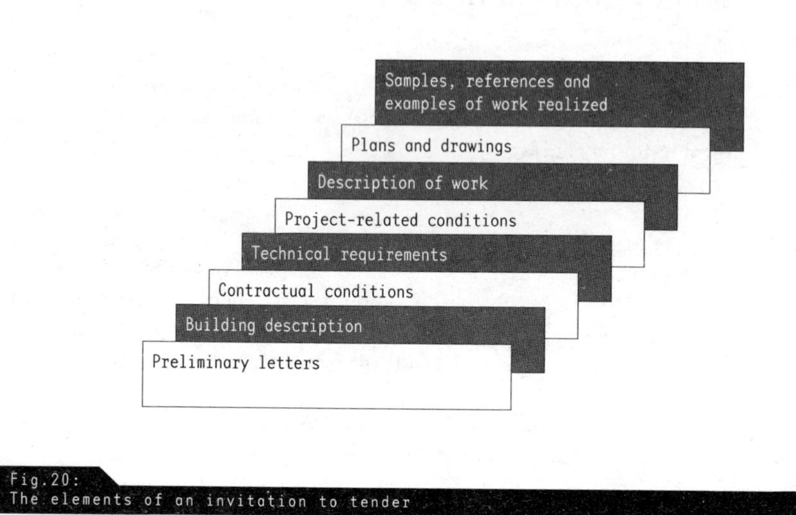

Fig.20:
The elements of an invitation to tender

	Textual elements	Drawing elements	Miscellaneous elements
Concrete elements	- Functional description - Detailed description of work - Records of award, negotiations and meetings - Building description - Preliminary remarks - Building site conditions - Reports - Special contractual conditions where applicable - Special technical require-ments where applicable	- Plans - Sketches	- Tests - Samples - Reference objects - Realized examples
Standardized elements	- Standard service text - General contractual conditions - Special contractual conditions - General technical require-ments - "Accepted technical rules" - List of manufacturers	- Reference drawings - Key details - Manufacturers' details	- Reference drawings - Key details - Manufacturers' details

Fig.21:
Systematizing the elements of an invitation to tender

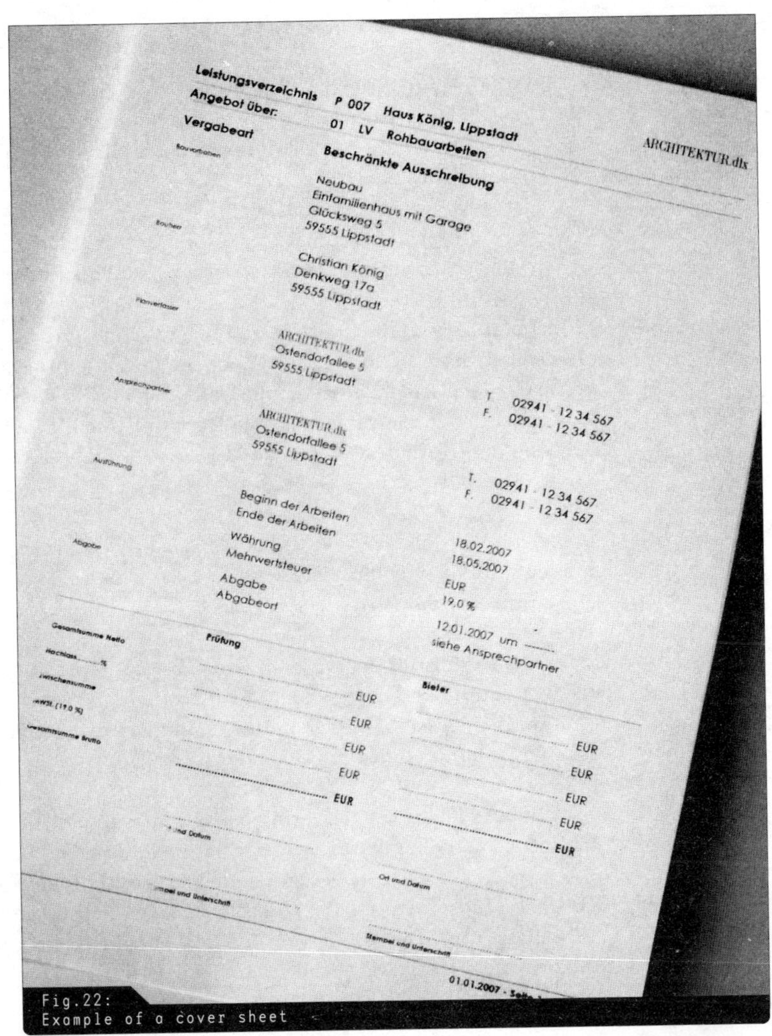

Leistungsverzeichnis P 007 Haus König, Lippstadt
Angebot über: 01 LV Rohbauarbeiten
Vergabeart Beschränkte Ausschreibung

ARCHITEKTUR.dlx

Bauvorhaben
Neubau
Einfamilienhaus mit Garage
Glücksweg 5
59555 Lippstadt

Bauherr
Christian König
Denkweg 17a
59555 Lippstadt

Planverfasser
ARCHITEKTUR.dlx
Ostendorfallee 5
59555 Lippstadt
T. 02941 - 12 34 567
F. 02941 - 12 34 567

Ansprechpartner
ARCHITEKTUR.dlx
Ostendorfallee 5
59555 Lippstadt
T. 02941 - 12 34 567
F. 02941 - 12 34 567

Ausführung
Beginn der Arbeiten 18.02.2007
Ende der Arbeiten 18.05.2007

Abgabe
Währung EUR
Mehrwertsteuer 19,0 %
Abgabe 12.01.2007 um ------
Abgabeort siehe Ansprechpartner

	Prüfung	Bieter
Gesamtsumme Netto	EUR	EUR
Nachlass ......%	EUR	EUR
Zwischensumme	EUR	EUR
MWSt. (19,0 %)	EUR	EUR
Gesamtsumme Brutto	EUR	EUR

Ort und Datum

Stempel und Unterschrift

01.01.2007 - Seite

Fig.22:
Example of a cover sheet

Cover sheet/accompanying letter

The cover sheet is both an introduction to and a summary of the contents of the invitation to tender. It includes brief summary of all the information required for the building firm to process the tender, along with the conditions of application. The cover sheet should contain the following information:

_ Details of the sender and recipient of the documents
_ Date
_ Definition of the building project
_ Location, nature and scope of the work required
_ General conditions relating to the building project (type of award, timescales etc.)
_ Express invitation to tender
_ Conditions of application
_ List of documents enclosed (contractual documents)

Description of the building project

Details about the location, nature and scope of the services required are needed, along with a definition of the building project, so that it can be clearly identified. These should be as succinct as possible. If more information about the project is required, this should be included in the building description. › Chapter Structuring service specifications, Approach to functional service specifications, Functional specification with designs

Information about the awarding procedure

The awarding procedure should be clearly defined in the cover sheet and must be given by public clients. › Chapter Organizing the tender, Timetabling the invitation to tender, Time invested by participants

Viewing dates

Identifying the location is also important in relation to possible viewing dates. Such dates should be identified on the cover sheet, giving location and date. The same applies to planned inspection dates in other documents not included in the tender package.

Tendering period/ submission

Details about the tendering period, the submission date and binding dates for the tender are equally important. › Chapter Organizing the tender, Timetabling the invitation to tender The tendering period identifies the date by which the bid must be submitted. The bidder will be committed to the offer made until the date given. Bidders should be informed about the essential award criteria (e.g. price) in the cover sheet for the tender documents.

Binding nature of the bid

To avoid misunderstanding, the cover sheet should include a formulation stating that the bidder will incur costs as a result of bidding. The formulation could be:

"We request your binding bid for this building project, at no cost to us or our client ..."

Conditions of application/ admissibility criteria

Planners can influence potential applicants through the application conditions in the accompanying letter. Possible conditions can regulate the use of subcontractors, or admit or exclude group bidding for the award.

\\Example:

"The … project relates to a three-storey office building with a gross floor area of approx. 10,000 m^2, to be constructed as a reinforced concrete skeleton structure with exposed concrete. The building plot is between … street and … street, and can be accessed only by the entrance at the junction of … street and … street. The precise location of the building site can be found in the site plan."

This example makes it clear how little the building description covers. It provides important information about the desired building method (skeleton structure with exposed concrete), the nature of the work (reinforced concrete building work), size (gross floor area, three storeys), function (office building) and the location of the building site (reference to the site plan).

Proof of suitability

Planners can use a request for proof of suitability and admission criteria to check whether bidding companies are qualified to carry out the work. As well as an informed check on suitability, by reference to projects previously carried out, for example, it is also possible to call for information about a company's economic situation, to ensure that it is sufficiently liquid. It is also customary to ask for information about the building firm's capacities, its membership of professional organizations and its liability insurance, stating the minimum sum covered.

Building description

The building description contains other general information about the building project. It provides the company carrying out the work with a general summary of the building project, with no detailed information about individual services. Bidders should complete the picture for themselves with construction descriptions and information about the key conditions affecting costs for the building project.

For larger building projects it can make sense to include more precise information about the body of the building or the structure and organization of individual building sections in the building description, so that bidders can form a clearer picture of the possible building phases.

Contractual conditions

The invitation to tender aims to prepare the way for a contractual relationship between the client and one or more companies realizing the project. Against this backdrop, provisions governing contractual modalities for carrying out the building work are important, as well a description of the services required.

Planners prepare the way for the future contract appropriately in the invitation to tender, by stating general and particular contract conditions.

\\Tip:
As the regulations involved are often very complicated, it is advisable to use recommended contract texts, drawn up by professional associations, for example, or at least to have lawyers draw up the contract conditions for larger projects.

General contract conditions are available in the form of complete sample contracts, but as a rule special contractual conditions must be formulated as business conditions laid down by the client.

General contractual conditions

General contractual conditions are based on national or international standards for managing building projects. They contain important information about:

- Nature and scope of the work required (details of contract elements and their ranking, and information about rights of change or extension relating to the building project)
- Compensation (provisions for dealing with claims for compensation in cases of deviation from the work as described)
- Implementation (provisions for supervision of the work by the client, for ensuring general order on the building site and the use of on-site facilities by the company carrying out the work; provisions governing rights of appeal if the company carrying out the work has complaints about a service required by the client or the planner)
- Implementation documents (information about handing over the documents relating to implementing the services required)
- Timings (general provision, for example ensuring that building work will begin within a stated period of time if no date was fixed contractually)
- Impediments (fixing procedures if impediments are in the offing. For example, obstructions should be notified to the client in advance and their effect described, so that counter-measures can be taken)
- Cancellation (provisions for cancellation by the client or the company carrying out the work)
- Liability (details of the contractual parties' responsibilities)
- Contract penalties (provisions governing modalities for contractual penalties not covering the penalty level)

- Acceptance (setting down timings for legal acceptance of building work)
- Guarantee (provisions for securing the client's claims after the building work has been accepted)
- Settlement (details about how and in what order settlement must take place after completion of work required, or parts of that work)
- Work paid by the hour (provisions for dealing with remuneration for services required that are not contained in the description of services, for example a commitment by the firm carrying out the work to inform the client before undertaking such work)
- Payments (general provisions governing instalments, part-final and final invoices, for example, timings are laid down for the duration of the final invoice check)
- Security (provisions governing mutual security for the contract partners, for example in the form of guarantees or security retentions)
- Disputes (provisions in case of dispute, such as fixing the client's location as the place of jurisdiction)

Special contractual conditions

Special contractual conditions can relate to the same matters as the general contractual conditions and complement them in certain points. They serve as an addition to the general contractual conditions, and not as a substitute for them. Typically, special contractual conditions are included in the tendering documents if there is already a provision in principle in the general contractual conditions. The following areas are also addressed:

- Invoices (invoices must be identified according to their purpose as instalment, part-final or final invoices, and always numbered continuously. Other formal requirements can deal with the sequence of the work carried out, identifying it according to the description of work required, for example.)
- Special payment modalities (provisions governing the client's payments to the company carrying out the work and the conditions to which the payments are linked. For example, a payment plan can be agreed, giving information about the level and date of payments. Payments are often agreed at particular times to cover the work carried out to this point.)
- Basis for establishing the price (the bidder's calculations used to determine the prices in the bid)
- Flexible price clauses for wages or materials (provisions allowing for contract prices to be modified if the agreed wage levels or building material prices change during the building phase)

- Notification of additional costs (provisions establishing that the client be informed at an early stage of any additional costs that may occur)
- Subcontractors (subcontractors are used to provide services that a company cannot itself cover. If the use of subcontractors is to be excluded or is permissible only under certain circumstances, this should be laid down in the special contractual conditions.)
- Competition restriction (inadmissible competition restrictions arise from prior agreements that are unfavourable to competition between bidders relating to the submission or non-submission of bids, to prices or profit supplements. Special contractual conditions lay down the consequences of behaviour that is unfavourable to competition.)
- Price reductions (are regularly agreed as a percentage and deducted from all invoices appropriately)
- Environmental protection (Normally no concrete environmental protection measures are formulated. It is customary for the special contractual conditions to refer to reduction of environmental damage by the building measures.)
- Changes to the contract (Contract alteration modalities should be stipulated in the special contractual conditions. For example, it can be agreed that alterations to the contract must be in writing.)

Technical requirements

As a rule, planning a building project and describing the work required to realize it end when a certain degree of detail has been reached. Everything else is fixed by the agreement on technical requirements. This contains instructions about the way the work is to be carried out. For example, planners might provide a drawing of a reinforced concrete wall, and possibly supplement it in the text with description containing details about formwork, reinforcement and concrete. But they will not describe in detail how the formwork should be constructed, the reinforcing steel secured in position or the concrete compacted. Such information forms part of the specialist knowledge of the company carrying out the work, and will be conveyed by the planners to the building firm via the technical requirements laid down in the invitation to tender.

Technical requirements are available as a comprehensive package of provisions for most services provided by different trades. ›Appendix They contain relevant stipulations for a large number of building jobs in the form of a minimum standard. Special technical requirements are formulated to define a higher standard.

General technical requirements >

General technical requirements are standards that apply in terms of the generally acknowledged rules for a particular technology.

Regulations are usually arranged specifically to trades and contain information about the sphere of validity, the substances and materials used, implementation, additional services that form part of the service as a whole, and about financial settlement and hints for compiling a description of the services.

Special technical contractual conditions

Special technical requirements are regulations that are used either to complement the general technical requirements or that apply to areas not previously regulated. For example, a special technical requirement can relate to a building process not covered by the general regulations, or can stipulate a higher dimension tolerance requirement to complement the existing minimum requirements.

Special technical requirements are based on standards, as well as on other sets of technical regulations, manufacturer's guidelines or provisions, and instructions from interested parties.

It is also possible to agree on more demanding requirements taking account of the current state of technology and of science and technology for certain services; these requirements will be based on individual licences.

Furthermore, there are special technical requirements for intermediate acceptance: for example, when for technical reasons certain pieces of work have to be accepted during the building period as they will be inaccessible at a later stage because of building progress.

\\Hint:
If certain provisions apply to one particular building project, they should be addressed in the contractual conditions relating to the project, and not in the special contractual conditions, which are usually formulated to cover several building projects.

\\Hint:
The generally acknowledged rules of technology are a set of regulations based on technologies that have proved their worth over a long period of time. A higher standard is set by a level of technology that represents the latest technical progress, but need not be tried and tested. A further step upwards is offered by a level of science and technology that takes the most recent scientific insights into account.

Project-related contractual conditions

Project-related information covers the general conditions of the building project. They cover all the regulations of a contractual and technical nature affecting the building project as such.

Information about the building site

These particular contractual conditions have to be compiled specially for every project. They should contain the following information about the building site:

_ Location (address and description of where the building site is situated)
_ Access (how to reach the building site)
_ Storage space (areas that will be at the contractor's disposal for work on the building programme)
_ Lifting equipment (lifting equipment such as cranes or hoists are often in place on building sites and can be used by various firms to transport their building materials)
_ Scaffolding (scaffolding may be placed at the disposal of other firms)
_ Connections for electricity, water and sewage (the appropriate supply points are fixed before building starts as part of the site equipment)
_ Sanitary facilities (if available)
_ Waste disposal
_ Telephone connections

Apportioning general building site costs

The general building site costs can be contractually apportioned to all the contractors involved. Costs for setting up site signs, using on-site equipment and waste disposal can also be apportioned in the project-related contractual conditions.

Implementation period/contract deadlines

Stipulations about the time available for the work are particularly important. All statements relating to this are fixed in relation to the project. They include statements about the beginning and end of the building work. These periods are binding for later implementation of the building commission, and if not observed they represent a breach of contract with the possible consequence of claims for damages, or a contract penalty. Only contractual periods that the company undertaking the work has acknowledged in the project-related contract conditions are legally binding. If intermediate deadlines other than the starting and finishing deadlines are agreed contractually with the company undertaking the work, these must be identified as individual fixed periods in the project-related contract conditions.

Contract penalty

If contract deadlines are not met, this usually means claims for damages by the client against the company undertaking the work. Here, only loses that have actually resulted can be considered. If other provisions are also made for handling breach of contract, these must be indicated appropriately in the project-related contract conditions with reference to a contract penalty.

It is also possible to agree on other project-related contractual conditions if required, for example stipulations about parallel services by other contractors, or provisions for clearing and cleaning the building site.

Tender specification

The tender specification is the key element in an invitation to tender. The distinction between functional and detailed tendering is based solely on the nature of the tender specification. A directory of services is used for a detailed tender specification, and a programme of services is drawn up for a functional tender specification. In exceptional cases, building descriptions are used as functional tender specifications. › Fig. 23

The procedure for drawing up a detailed or functional tender specification is discussed in detail in the final chapter.

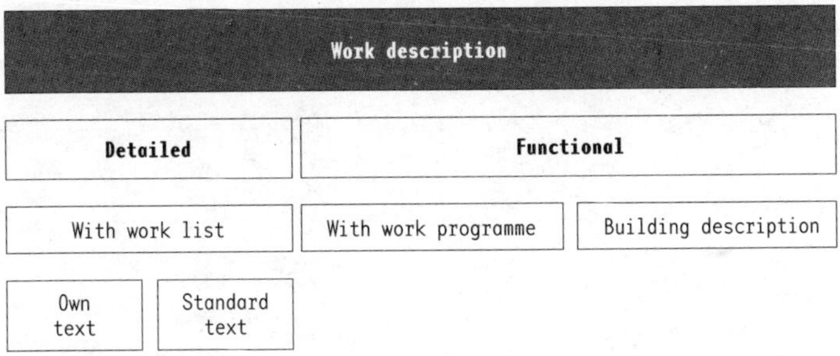

Work description

Detailed | Functional

With work list | With work programme | Building description

Own text | Standard text

Fig. 23:
Tender specification types

DRAWING ELEMENTS

The plans, drawings or sketches appended to an invitation to tender should make it easer for the contractors to compile their bid in terms of the services to be calculated. They will therefore need all the planning documents necessary for general geometrical orientation, and for understanding the services required.

Range of drawn descriptive elements

The spectrum of drawn descriptive elements extends from simple hand-drawn sketches to technical drawings, with plan content varying from site plans to scaled implementation details. › Figs. 24, 25

References

The architect can use references to particular planning details in the tender specification to identify particular points that do not emerge directly from the descriptive text or that are more easily conveyed by a drawing.

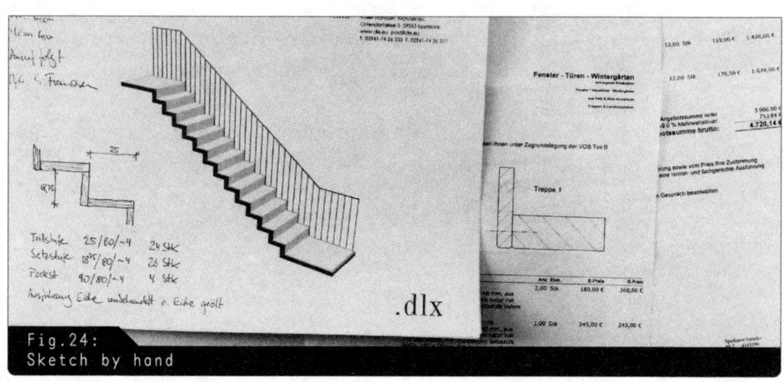

Fig.24: Sketch by hand

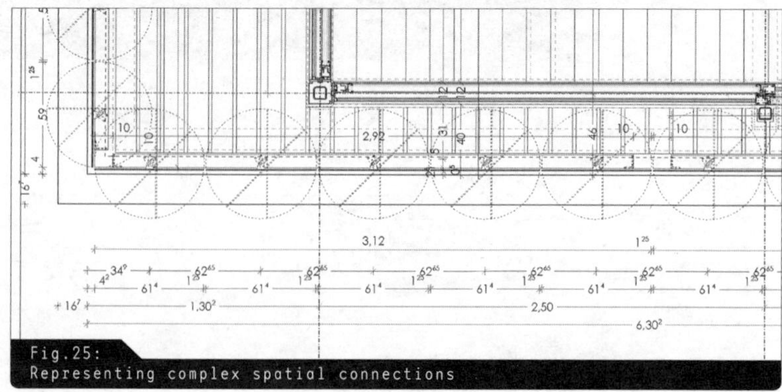

Fig.25: Representing complex spatial connections

\\Hint:
Building work is done on the basis of the architect's working plans, not of the tender specification. The tender specification forms the basis for compiling the tender bid. If information in the tender specification contradicts the plans, each case should be examined individually, following the contractually fixed hierarchy where applicable.

OTHER DESCRIPTIVE ELEMENTS

Specimens

If texts and drawings cannot define the services required adequately, specimens can be used. For example, if the reinforced concrete requires a particular surface structure that is not covered by the general regulations and definitions for surface quality in exposed concrete, it makes sense to construct a specimen surface and make this accessible to bidders or even enclose it with the invitation to tender (e.g. a related wood veneer that is already in the building).

If a specimen is available, the invitation to tender text need not embark on a full description, but concludes with the formulation *"Finish as per specimen ..."*

Markings

Markings can also be addressed in the invitation to tender. Markings need to be considered in the case of materials such as natural stone, as the appearance of stone can vary considerably. Here, verbal descriptions and drawings are almost impossible. The company carrying out the work should be required to lay a representative area of the material, so that the crucial criteria such as colour, shade and the nature and distribution of inclusions can be fixed according to the sample.

It is also possible to set up show rooms in which the client can see and assess the effect made, from surface materials to individual pieces of furniture and fittings in context.

Reference items, examples of finish

Reference items are particularly important when building in existing stock, or in the case of ensembles. For example, when designing the exterior of a building, the choice of brick can be specified to be the same as the existing buildings, with the same pattern of joints, without the planner needing to explain format, colouring or bond. Reference to buildings that are already in existence and the quality achieved there can also form the basis for describing building work.

STRUCTURING SERVICE SPECIFICATIONS

We distinguish between three basic starting situations for functional or detailed service specifications. › Fig. 26

1. No design is supplied
2. Plans or planning permission are available
3. The final planning stage has been reached

APPROACH TO FUNCTIONAL SERVICE SPECIFICATIONS

Aims

The aim of functional tender specifications is to bring all the necessary requirements for a building together.

Functional specification instruments

Drawing up a functional tender specification can be considerably facilitated by recourse to various descriptive instruments. They include:

_ Building descriptions
_ Building programmes
_ Room programmes
_ List specifying all the work required

Construction and fittings and furnishings manuals are a further step in relation to detail in tender specifications. Their language is not essentially functional. It contains too many concrete requirements, and thus runs counter to the open concept principle of a functional invitation to tender. But such manuals can be used as part of a functional invitation

	Detailed tender specification	Functional tender specification	
Planning stage	With working plans	With design	Without design
Ways of examining the bid	Financial approach (bid price)	Design, functional, technical and financial approach	Technical and fincancial approach

Fig. 26: Service specification characteristics

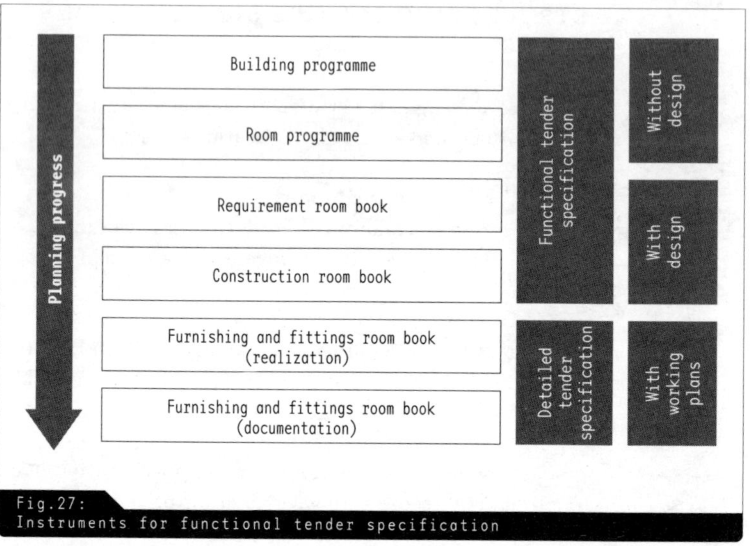

Fig.27:
Instruments for functional tender specification

to tender if a quality requirement is fixed beyond further discussion and is to be implemented when furnishing or decorating certain rooms, for example. › Fig. 27

The instruments listed either serve as a basis for drawing up a functional tender specification, or become part of the specification themselves. Without a design, only a building or room allocation programme can be drawn up; a design must be made available if a book listing all the work required for the rooms ("room book") is to be compiled.

Whether a design is available or not, nothing changes the basic principles of functional tender specification. The aim is to define everything

\\Hint:
Building programme and room allocation programme can be used for functional tender specification even when a design is provided, as long as they do not contradict it. But it is customary to provide a more detailed description of services required in terms of rooms, structural elements and construction products.

the client requires, although more creative input is required from the bidder if no design is provided.

Functional tender specifications without a design

A building programme and a room allocation programme, offered without a design, simply describe requirements for the building as a whole, for individual parts of a building or for areas intended for a particular use. The bidder is responsible for design, technical, use-oriented and economic planning.

No.	Field	Requirement
I	Area	Sample street 12, 00001 Sample town
II	Requirement (description)	Inner-city office complex
III	Nature of project	Conversions
IV	Use	Office, canteen
V	Plot size	10,000 m^2
VI	Number of floors	3
VII	Cellars	yes (1 floor)
VIII	Building structure	2 Main building office 1 Ancillary building, canteen
IX	Office space	from ... m^2 to ... m^2
X	Canteen space	from ... m^2 to ... m^2
XI	Office type	Individual offices and open-plan office
XII	Individual offices	from ... m^2 to ... m^2
XIII	Open-plan office	from ... m^2 to ... m^2
XIV	Access	The building is to be connected to public utilities and transport
XV	Parking	Underground car park in cellar, parking spaces on the north side of the building, 1 parking space per workspace
XVI	Waste	Central waste disposal
XVII	Open spaces	Park with pond
XVIII	Rules and regulations	Development plan, local building requirements

Fig.28:
Sample structure for a building programme

Building programme

A building programme makes basic statements about a building. It first gives details of the property, for example covering use, office size, office type, the number of floors, number of offices per floor, or cellarage.

Building programmes contain information about the building project in general, supplying additional information about connections to public services (sewerage, water, gas, electricity and telecommunications), the transport system and access to outside areas.

Building programmes must specify requirements for the prescribed areas of use. These requirements can be differentiated and concretized in part. For example, it is possible even at this stage to fix sound insulation requirements for an area with individual offices. › Fig. 28

Room allocation programme

A room allocation programme provides a more refined definition of the requirements. It will give information about rooms and use areas, and also about how they are placed and linked together. The bandwidth of possible information in a room allocation programme depends on the planning stage reached. Sensible subdivision of the areas according to the following criteria forms a good basis for a description system:

_ Use
_ Number
_ Size
_ Position and orientation › Fig. 29

Function scheme

A function scheme shows how individual areas relate to each other without illustrating the areas required. Essential links between individual areas are indicated in order to clarify the sequence of events arising from a particular use. › Fig. 30

Graphic room allocation programme

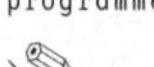

Information from a tabular room programme and function scheme can be summarized in a graphic room allocation programme. This can contain basic elements of the architectural design that have already been

\\Tip:
Symbols can be used to add further information, such as data about the required lighting (daylight/artificial light), to graphically presented room programmes.

Use	Number	Size	Position and orientation
Reception foyer	1	150 m^2	Ground floor north side/west side
Canteen	1	200 m^2	Ground floor north side/east side
Kitchen	1	80 m^2	Ground floor centre/east side
Office	4	each 25 m^2	Ground floor south side/west side
Events room	3	1 x 200 m^2 2 x 50 m^2	Ground floor south side/east side

Fig.29:
Example of a tabular room allocation programme

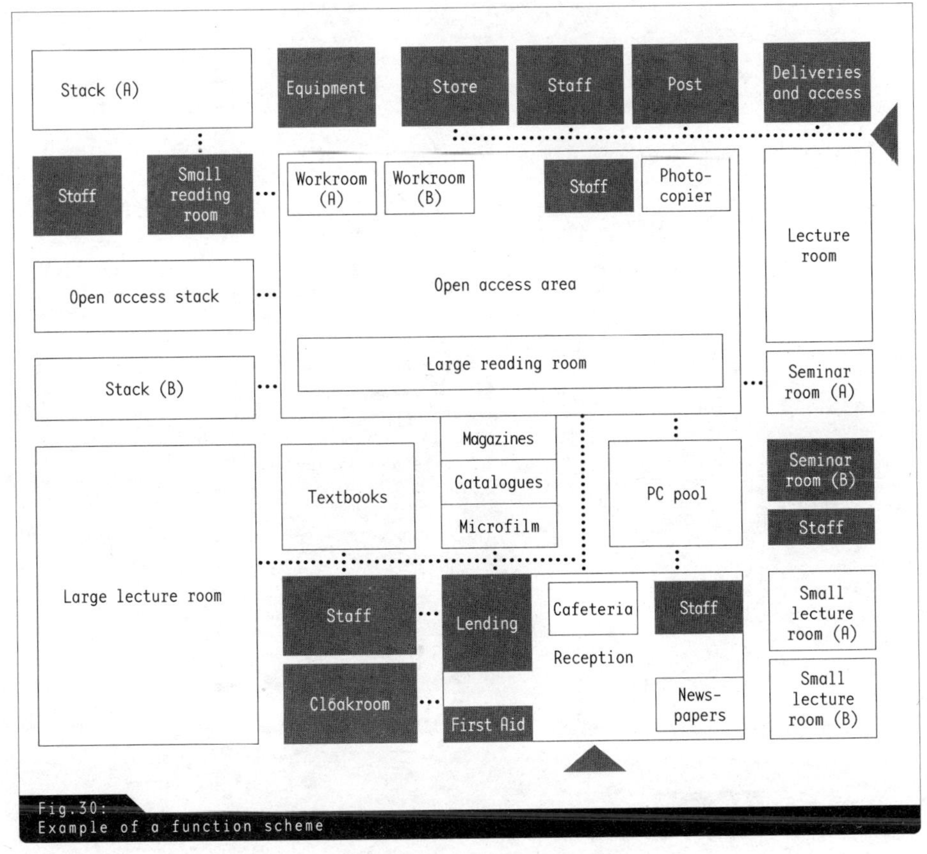

Fig.30:
Example of a function scheme

determined: formal statements are presented, taking the areas and rooms required and the way they relate to each other into account. › Fig. 31

Other means of description are suitable as well as building and room allocation programmes. A building description for drawing up a functional tender specification without a design is one of these.

Building description

The building description essentially provides a rough idea of the building project as a whole, but can also convey functional details. › Chapter Structuring an invitation to tender, Textual elements, General information about the building project Unlike the room allocation programme, which is based on spatial organization, a building description is structured in terms of construction or trades. This is also why it is only very roughly suitable as a basis for functional tender specification. Saying "reinforced concrete hall, area 2000 m^2" is a very rough description, but it is perfectly appropriate for use

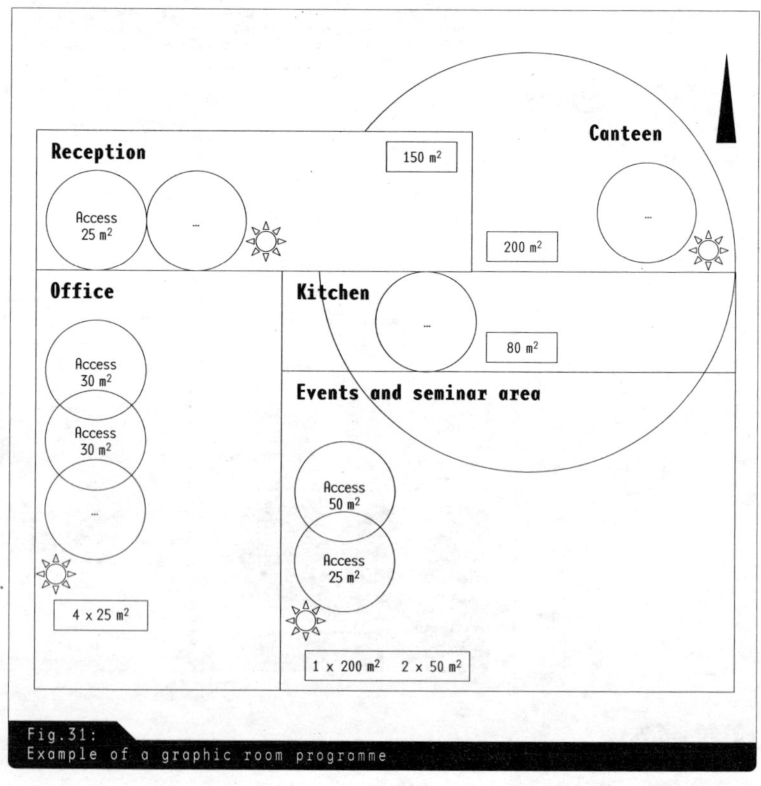

Fig. 31: Example of a graphic room programme

as an element in a functional description. Specifying "reinforced concrete" rules out alternative construction elements, such as steel girders.

The more specifically the building description addresses a particular construction method, the more effectively unwanted elements will be ruled out as alternatives for bidders.

Functional specification with design

If a design is available, definite requirements for the building project are already in place, and these are shown in plans. But the functional character of a tender specification with design supplied is still guaranteed, as it remains possible to define the quality of individual construction systems, structural components or construction products. In this context, using a room book – a set list of all the services required to complete a particular room – is a helpful device for systematizing requirements.

Room book

A room allocation programme can be addressed in greater detail by introducing room books at an appropriate point in the planning phase. This is a system that makes it possible to supply information about any planned room. A room book sums up the space required systematically and defines use requirements. Each room book should contain the following information to ensure unambiguous identification and simplify further use of the information provided:

_ Room number, following a defined system for structuring the building project
_ Definition of the room
_ Information about the nature of the area
_ Information about requirement or furnishing characteristics

\\Hint:
Room books can be used for different purposes: identifying planning needs for a design; forming the basis of a function tender specification; as sales documents for marketing buildings; supporting the site management teams during the construction phase; recording the state of affairs when work is completed to provide a guarantee; or in relation to running the building. They also provide a useful basis for planning as a stocktaking device for future extension or conversion measures.

Room book types

There are three different kinds of room book, fulfilling different purposes and so requiring different planning levels:

_ Requirement room books
_ Construction room books
_ Furnishing and fittings room books

Room book

Requirement room book		**Sheet:**	05
Building project:	Weinreich Versicherungen, Musterstadt South (P45/145)		
Building type:	Office complex		

Date:	05.03.2007	**Prepared by:**	Mr Müller
		Approved:	Ms Sanders

Room description:	Office	**Floor:**	1st floor
Room number:	1.304		

Technical requirements

Area	**Requirement**	**Statistics**
Statics ...	Maximum deflection ...	f = 1/300 ...
Building science ...	Fire prevention as per DIN 4102 ...	Structural component at least F30 ...

Requirements by function

Area	**Requirement**	**Statistics**
...	...	...

Design requirements

Area	**Requirement**	**Statistics**
...	...	...

Financial requirements

Area	**Requirement**	**Statistics**
...	...	...

Ecological requirements

Area	**Requirement**	**Statistics**
...	...	...

Fig.32:
Example of a requirement room book

Requirement room book

A requirement room book plays a key part for a functional tender specification with design supplied. A sheet with a table of all the known requirements is drawn up for every room or area. › Fig. 32

Construction room book

A construction room book offers another form of description, providing a detailed description of the construction, but not the fittings and furnishings in a particular room. › Fig. 33

Room book

		Page:	05
☐ **Construction room book**		**Building project:**	Oberstrasse 1 12345 Dorla
☒ **Furniture and fittings room book**		**Building type:**	

Date:	05.03.2007	**Prepared by:**	Mr Müller
		Approved:	Ms Sanders

Room description:	Office	**Room height:**	3.00 m
Room number:	1.304	**Area:**	20.60 m^2
Floor:	1st floor	**Type:**	Use

No.	Element	Fixtures/Structure	Properties	Quantity
1	Floor	Reinforced concrete slab in site-poured concrete Impact sound insulation Separating layer Screed, carpet	C20/25 PE sheet ZE 20, d = 50 mm	1
2	Ceiling	Reinforced concrete slab in site-poured concrete False plasterboard ceiling Grouted joints Paint ...	C20/25	...
3	Wall	...	...	...
4	Window	...	...	...
5	Doors	...	...	...
6	Lighting	...	...	...
7	Power supply	...	...	...
8	Heating	...	...	...
9	Ventilation	...	...	...

Fig. 33:
Example of a furnishing and fitting and construction room book

\\Hint:
A furnishing and fittings room book giving detailed information about every room is suitable primarily for preparing a detailed tender specification with a complete list of technical and general requirements for each trade. Care should be taken to ensure that the services are listed in relation to trades, and not in relation to rooms.

Furnishing and fittings room book

A furnishing and fittings room book simply provides a complete description of the furniture and fittings for every room. It lays down which elements are to be fitted in each room, and in what quality and quantity. Each element is listed by number and manufacturer's description, or in comparable detail.

Structuring a functional tender specification

Building breakdown

The building should be broken down in order to systematize a tender specification by function offering a general programme of services – in contrast with a trade-oriented tender specification with a complete list of services. Rooms are recorded clearly and systematically, and numbered consecutively, following their position in a part of the building and a storey.

For further breakdown, a list of use or function areas should be drawn up. At this level, it is already appropriate to provide information

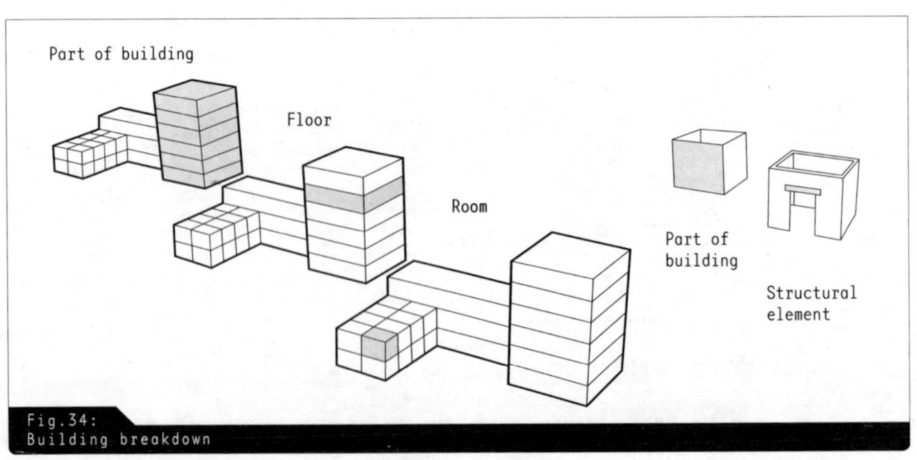

Fig.34:
Building breakdown

about individual supply and technology elements, foundations, the loadbearing structure, the façade and the roof. If individual uses are known or intended, requirements can even be defined on the basis of structural components, such as a non-loadbearing wall between two offices. › Fig. 34

A building project needs to be broken down only to a certain extent: it must be possible to define different requirements on the basis of the client's

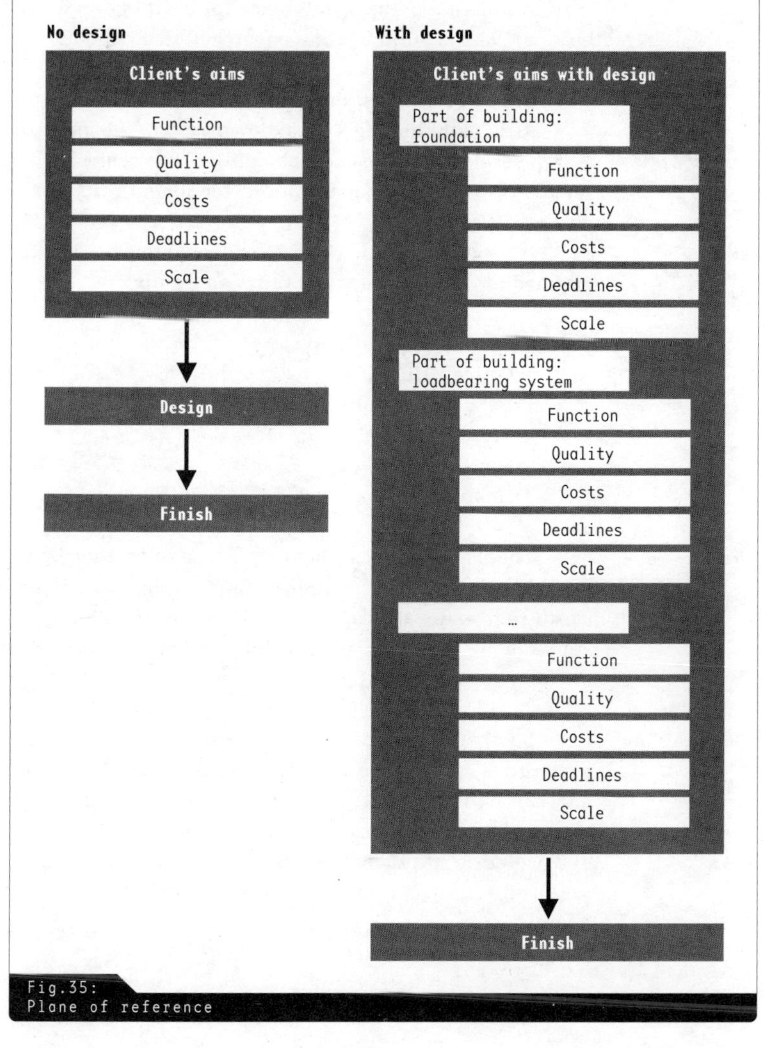

Fig.35:
Plane of reference

intentions and the known general conditions. If the client prescribes a function specifically, it does not make sense, and it is usually not possible, to break such a function down into its individual components. ›Fig. 35

Breakdown by building or room allocation programme and by room book

The above-mentioned building programme, room allocation programme and room book are structured to provide a breakdown. They can be used in relation to the different breakdown levels as a function tendering specification with programme of services, by allocating requirements. They do not have to be used, however. All that is fundamentally necessary is to establish a plane of reference for defining requirements. This could be a part of the building, or a construction component.

Drawing up a requirement profile

After establishing a breakdown system the use profile (client's aims and general conditions) can be applied successively to the smallest unit selected (a room or a construction component).

Rough description of the function of a building

It is recommended that the first step should be to allocate the planned building project to a function group:

_ Housing construction
_ Office building
_ Department store
_ School, college
_ Factory
_ Hospital

Requirement categories and aspects

The use profile can be differentiated further by identifying design (social and aesthetic), technical, functional, financial, and other categories if required. Each category will contain coherent requirement aspects. For example, building science and construction are both requirement aspects within the technical requirement category.

Individual requirements and ratings

Further subdivision is also possible on the basis of individual aspects. These again refer to certain subsections of the requirement aspects, and can be fixed more precisely in terms of ratings. Individual requirements regularly contain references to standards and regulations laying down certain minimum values. One possible individual requirement in terms of building science is fire protection for a door that can be fixed at a minimum rating of F30.

The individual requirements and the ratings that lie behind them can be compiled clearly using the breakdown system shown in Fig. 36.

Requirement category	Requirement aspect	Individual requirement	Statistics
e.g. Technical requirements	e.g. Building science	e.g. Fire prevention	e.g. F60
			...
		...	...
			...
	e.g. Statics	...	...
			...
		...	...
			...
e.g. Aesthetic requirements	...	...	...
...	...	...	...

Fig.36:
Example of a system for recording service programme requirements

Possible ways of presenting a functional tender specification

However, summing up requirements in a service programme is not the only descriptive language available. Requirements can also be defined in continuous prose, so long as this retains concrete allocation of the requirement to a particular element (e.g. a part of the building or a room).

Defining requirements

Requirements should be defined in full and unambiguously with a view to the client's wishes and the general conditions of the building project. Possible requirements are explained in greater detail with reference to this below, following the categories identified above.

\\ Important:
Considerable variations are possible in the depth of analysis required to describe a client's intentions. It is possible that a client will simply identify an output value for a production plant. It is then up to the bidder to investigate all other criteria within the general conditions, which cannot be changed. If this procedure is being followed, it is neither possible nor appropriate to break the building project down.

\\ Hint:
Ratings provide a clear, measurable basis for individual requirements. If no ratings are given, or if they are defined only qualitatively (e.g. enhanced sound insulation), this may give undesirable scope for the bidder's interpretation.

Design requirements

The design requirement category includes both aesthetic and social aspects. The content of this category is largely a matter of the client's sensibilities, and includes aspects like convenience, privacy and comfort in the social sphere, and architectural quality, elegance and prestige in the aesthetic sphere.

These subjective requirements might include specifying high quality building materials or imposing public areas (for example a spacious atrium).

Function-oriented requirements

Functional requirements are also determined by the client's aims. Mere allocation to a general function group (e.g. school) identifies key features of the intended function. The requirement aspects within this category provide information about the function grid, ceiling spans, the number of floors, floor area, usable area, variability for the ground plan, or possible changes of use for the building. Various requirements of a technical nature also follow from the building's function.

Technical requirements

Technical requirements are derived directly from the function, from standards and regulations, or from the client's express wishes. Essentially, all the areas relating to loadbearing capacity, stability and building science (e.g. heat and sound insulation, fire protection and waterproofing) are defined more fully by requirements.

For example, a technical requirement may follow from a client's wish for air-conditioned office spaces, and would have to be accounted for in the functional specifications.

Financial requirements

Financial aspects applying to a building are also largely determined by the client's aims. Aspects such as investment costs, building maintenance costs, running costs or yields from use are directly linked to the client's intentions and the function and technology of the planned building. One of the client's strategic aims could be to use alternative rather than fossil fuels in his or her building, which could mean higher investment costs, but lead to savings in the long run.

Ecological requirements

Themes like recycling potential, environmental soundness of the building materials used or implementation of an environmentally friendly energy concept are covered in the ecological requirement category. The general conditions for this category are primarily of a legal nature, but they can also be determined by the client's aims. The client may view state subsidies for environmentally friendly technologies as a reason for using them. However, the client may ask for a low-energy building without going

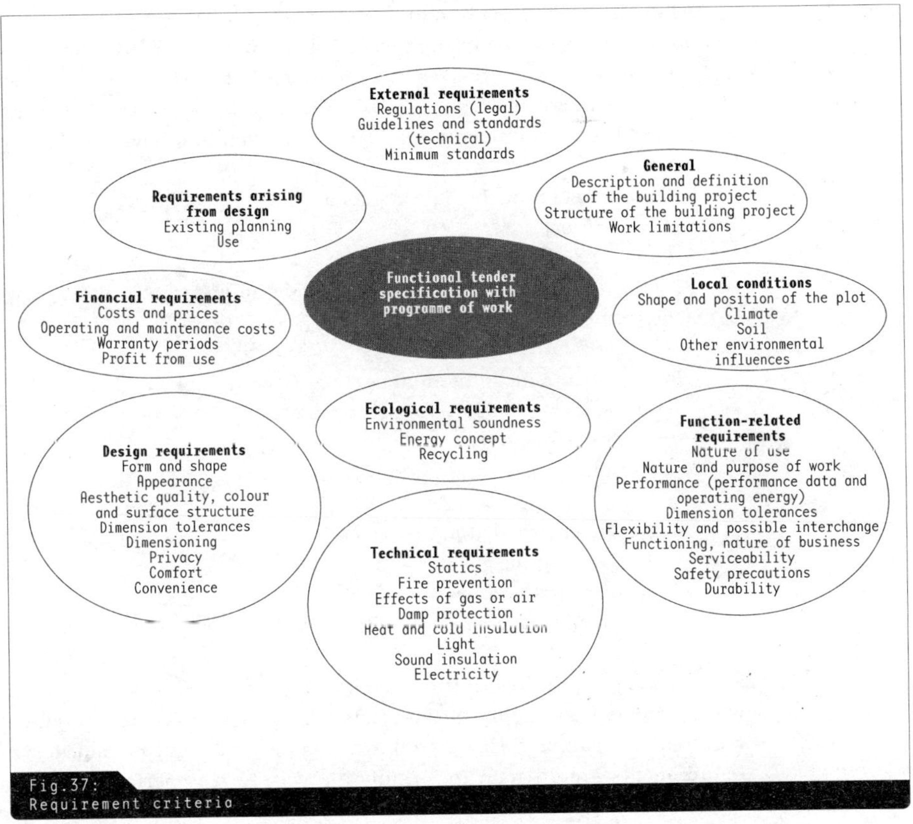

Fig.37:
Requirement criteria

into any further detail. Other examples of ecological requirements arise from long-term considerations in terms of pollution-free conversion or demolition of the planned building.

Other information for a functional tender specification

Other information not contained in the programme of services is important for a functional tender specification over and above the requirements that have already been mentioned. It includes details of legal and technical regulations, a rough description of the building, from which it should be possible to discover the local conditions for the project, and establish which works do not have to be included in the tender bid.

Figure 37 sums up the most important points to illustrate the possible requirements.

Requirements must be clearly presented in the programme of works. The presentation can be in the form of continuous prose, lists or tables. This largely depends on the way the programme of works is broken down. If the programme runs from the building as a whole to the individual room, a table of requirements can be systematized as follows:

Requirements for the building as a whole:
- Two storeys
- Solid construction
- Minimum requirement low-energy building in accordance with current regulations

Requirements for individual parts of the building:
- Conservatory finish with overhead glazing
- Natural ventilation
- Maximum summer temperature in rooms: 29 °C

Requirements for individual rooms:
- Study facing the garden
- Staircase with natural lighting
- Bathroom with separate toilet

Systematic listing of this kind quickly produces a programme of works that is very highly differentiated in individual areas. Individual requirements can be formulated down to a particular rating, but design requirements are difficult to formulate precisely. It is perfectly clear to stipulate that all the rooms in a hotel must have a sea view, but descriptions like "comfortable atmosphere" or "lounge style" are very much governed by individual ideas and experience, and serve little purpose unless they are not made more specific.

PROCEDURES FOR A DETAILED DESCRIPTION OF WORKS

Aims

The most economical bid for required work can be compiled with a detailed description of works. The tendering pathway must be described to bidders in detail, and in full, on the basis of the completed working plans. A tried-and-tested system is available to planners.

Structure

The tender specification is a general structural system that makes it possible to record individual works coherently. The individual works are itemized in tables, giving quantities by batch, trade and title. In practice, the tender specification breakdown shows the work to be done expressed by location and specialist fields. › Fig. 38

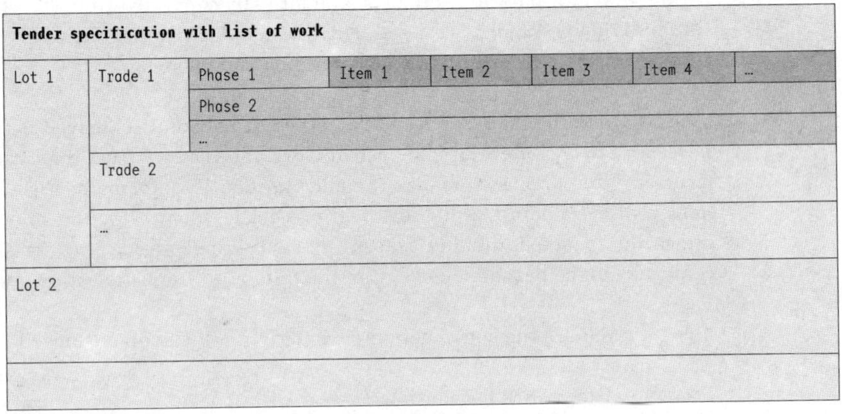

Tender specification with list of work							
Lot 1	Trade 1	Phase 1	Item 1	Item 2	Item 3	Item 4	...
		Phase 2					
		...					
	Trade 2						
	...						
Lot 2							

Fig.38:
Breakdown system for a tender specification

Subdivision by location or room is familiar from the procedure for a function tender specification. Subdivision by trades is another possible way of systematizing the work required.

Tender specification and unit price contract

The principle behind the tender specification is that it aims to make a direct enquiry about prices for the smallest descriptive elements, item by item. Bidders should provide unit prices for the items, which are described precisely by nature, quantity and quality.

The total price for an item is arrived at by multiplying the planned quantities by the unit price in each case. The sum of all the totals is the net total for the bid. When invoicing work after it has been carried out, the unit prices are quoted, but not the planned quantities. A unit price contract of this type is usually billed in terms of the quantities actually used.

\\Hint:
Unit prices are prices based on a unit, for example EUR 10/m^2.

Drawing up a tender specification

If a description of the work required is needed with a tender specification, the information has to be systematically converted from the plans into individual subjobs arranged by trade, as plans are, by their very nature, structured according to construction components.

If individual subjobs are to be listed with absolute clarity, it is first necessary to consider the sequence of work and the construction method. Adopting this approach makes it easier to identify subjobs of the same nature in more detail, and allot them to a trade. Identifying the subjobs, allocating them to individual traces and also actually describing the work required can be based on answering the following simple questions:

- What are the construction elements for which subjobs have to be described?
  Ceiling; wall; foundations ...
- What are the construction types for which subjobs have to be described?
  Masonry; reinforced concrete ...
- What are the processes for which subjobs have to be described
  Earth moving; reinforced concrete work ...
- What connections are there between the building phases and the trades?
  Excavations = earth moving; foundations = reinforced concrete work ...
- What subjobs can be allocated to particular trades and construction components?
  Reinforced concrete ceiling = formwork, reinforcement, concrete ...

The next step is to record more detail about the individual subjobs. Here we recommend that the relevant standards, guidelines and regulations are summed up, to provide a frame of reference for describing each subjob

\\Tip:
A useful instrument for compiling a tender specification with list of works is a room book detailing furnishings and fittings. This gives the number of rooms with a detailed description, as well as information about areas. Certain jobs can thus be recorded quickly and systematically in terms of quality and quantity for further description in the tender specification.

\\Hint:
Information about professional execution of building work can be found in the trading standards. These standards also contain information about classifying the work, the building materials and construction components used, the units to be used as a basis, the appropriate subjobs, and for invoicing and drawing up the tender specification (see Appendix).

in detail in terms of a sound, expert source of information about building materials, construction components for listing the works required.

The tender specification for all jobs can be drawn up stage by stage on this basis.

Lot

The lot is a complete award unit allocated to a company. Lots are to be seen as independent subprojects that can be defined equally on the basis of criteria relating to spaces (part lot) or to expert services (specialist lot, trade).

Part lot

Subdivision into lots by area usually only takes place for large building projects, and would allow for dividing a road-building project up into several phases or street construction contract sections. If the client perhaps intends to commission only part of the building work and allocate subsequent work to other firms, he or she must draw up appropriate lots.

Specialist lot

If the client is awarding the contract by breaking down trades, the term specialist lot is used. › Chapter Structuring an invitation to tender A trade can be split up into several specialist lots. For example, one metalworker can be commissioned to make railings and another to work on the façade.

Title and subtitle

Title

Titles are a further way of breaking down the building project below the level of the project as a whole. A title describes a part of a building or a particular trade within a lot or an overall building project. It can describe a subjob within a trade, without representing a unit that is complete in itself with its own price within the bid. The function of the title, as opposed to the lot, is to sum individual job items up in coherent sections.

\\Hint:
Just like an independent building project, lots can also be further broken down in terms of areas and expertise to relate to parts of a building or individual trades. Lots can also be defined on the plane of individual trades or titles, and are then awarded as a complete package of works.

Bundling individual jobs (items) that belong together in terms of speciality or physical area provides a suitable basis for establishing prices by placing an item within the overall context.

For example, the "metalwork" tender specification may contain titles such as "stairs and banisters", "doors and frames", and "fencing", in order to break the work down into coherent sections. Further differentiation can then be done in subtitles, such as "outdoor stairs and banisters" and "indoor stairs and banisters". Work can also be divided up according to individual structural elements. The tender specification for "shell construction work" can be broken down into titles such as "foundations", "floor slab", "exterior masonry", "interior masonry", "ceilings" etc., or even more fully in relation to the place where the work is to be done, as in "kitchen tiles", "toilet tiles" and then again into "floor tiles" and "wall tiles".

The number of breakdown levels is up to planners. They should break the tender down only to the extent that the complexity of the project requires. Breakdown by title and subtitle should always aim to form coherent units. As well as better understanding by making it easier to allocate the individual job items, this means that when comparing bids it is possible to compare something other than just individual items within the total price. In addition, planners can compare the bids in terms of the titles. For the above-mentioned example of metalwork, it could turn out when comparing the individual titles that one metalworker is offering the best prices for stairs and doors, but is bidding well above the average for fencing work.

Subtitle

Individual titles can be further broken down by the use of subtitles. For example, the reinforced concrete work required for a particular job can be summed up under a title, and the formwork and reinforcing material it requires in subtitles.

The extent to which there is differentiation between titles and subtitles or other breakdown levels (main titles where appropriate), and the sequence in which subdivisions are made in terms of working area and

\\Tip:
Listing by title makes it easier to evaluate bids for individual sections or trades. To do this, the total prices for items under a particular title are summed up and presented in a list of the individual titles.

- Earthworks
- Drilling
- Preparatory work
- Ramming, sieving, compressing
- Waterpipes
- Sewerage drains
- Draining
- Spray concrete
- Road and path construction
- Landscaping
- Injection spraying
- Underground cabling
- Rail construction work
- Masonry
- Concreting
- Natural stonework
- Artificial stonework
- Carpentry and woodwork
- Steel construction
- Sealing
- Roof covering and roof sealing
- Plumbing
- Dry construction work
- Composite heat insulation systems
- Concrete maintenance
- Plastering and stucco
- Ventilated curtain facades
- Tiling and slab installation
- Screed work
- Poured asphalt work
- Joinery
- Parquet work
- Metal fittings
- Blinds
- Metalwork
- Glazing
- Painting, varnishing, coating
- Corrosion prevention for steel
- Floor coverings
- Wallpapering
- Timber flooring
- Ventilation installations
- Heating and central water-heating facilities
- Gas, water and drainage facilities
- Low- and medium-frequency equipment
- Lightning protection
- Conveyor systems, lifts, escalators
- Building automation
- Scaffolding
- Demolition and dismantling

Fig.39:
List of different trades

expert fields depends on the size and complexity of the individual building project and the nature of the contract relating to it. Figure 39 shows possibilities of breakdown that are already available on the basis of specialist allocation of individual services.

Items

A single item within a tender specification is the smallest tender unit and represents a subjob within the building work. It is made up of individual descriptive elements defining the work to be done and the particular service required clearly and unambiguously. The descriptive elements can be formulated as required or put together from standard catalogues.

It is possible to include more than one activity within a particular item, as long as they can be seen to be the same in their technical nature, and for price calculation.

No.	Text	Item	Quantity	Unit	UP	TP
01.02.02.0001	...Formwork Floor slab...		50	m		

Fig.40:
Components of a tender item

Components of a tender item

Against this backdrop we recommend describing all the items systematically in a tender specification. In this context, the typical components of a tender item are grouped within the following categories:

- No. number
- Text descriptive text (short and long text)
- ITy item type
- Quantity the quantity worked out from the plans in terms of UQ
- UQ unit of quantity
- UP unit price (price for a unit)
- TP total price per item (unit price × planned quantity)

These categories are applied to every item, thus producing a specification arranged by lots, trades and titles or subtitles. The bidder enters the unit and total prices. › Fig. 40

The following information should be given in the individual categories for a tender position:

Number

The number helps to make it easier to find one's way around a tender specification. Each piece of work (subjob) in the same category in terms of techniques and pricing is identified by a particular number according to a defined breakdown key. This number relates directly to the way in which the project is broken down, and reflects this in the tender specification. In a relatively simple project a subjob can be identified by number as follows.

Lot	Trade	Title	Item	Index
01	01	01	0001	a

Index

The index can be used to show the relationship between a basic and an alternative item in the numbered list.

Numbering is consecutive at every level. › Fig. 41

Tender specification text

The tender specification text should be drawn up in long and short form by the planner inviting tenders.

01.02.01.0001	Activity 1 of title 1 in trade 2 of lot 1
01.02.01.0002	Activity 2 of title 1 in trade 2 of lot 1
01.02.01.0003	Activity 3 of title 1 in trade 2 of lot 1
01.02.02.0001	Activity 1 of **title 2** in trade 2 of lot 1
01.02.02.0002	Activity 2 of **title 2** in trade 2 of lot 1
01.02.02.0003	Activity 3 of **title 2** in trade 2 of lot 1
01.02.02.0004	Activity 4 of **title 2** in trade 2 of lot 1
01.03.01.0001	Activity 1 of title 1 in **trade 3** of lot 1
01.03.01.0002	Activity 2 of title 1 in **trade 3** of lot 1
01.03.01.0003	Activity 3 of title 1 in **trade 3** of lot 1
01.03.02.0001	Activity 1 of **title 2** in trade 3 of lot 1
01.03.02.0002	Activity 2 of **title 2** in trade 3 of lot 1
01.03.02.0003	Activity 3 of **title 2** in trade 3 of lot 1

Fig.41:
Example of the use of numbering

Short text

The short text is used essentially as a short textual summary of the service required for further use in drawing up the bid and raising the invoice. The short form must not lead to possible confusion between items, each item should be identified unambiguously: "masonry 36.5 cm, cellar", "masonry 36.5 cm ground floor", "masonry 17.5 cm ground floor" etc.

Long text

In contrast, the long text should describe the work required unambiguously and exhaustively, so that all the bidders understand exactly the same thing.

Freely formulated or standard texts

In principle, texts can be freely formulated, taking legal and technical regulations into consideration. But to simplify this process, planners can make use of standardized sample texts, which are generally available.

Standard texts

Such standard works catalogues are collections of texts that can be used to described work required or subcategories of it. These are structured according to predefined patterns and contain a range of information about building work, building materials, dimensions and units of quantity for various works.

The great advantage of standardized specification texts is that they mean the same to all bidders, and so no there is no unnecessary effort or additional risk when fixing prices. Furthermore, the usually modular

\\Tip:
Standardized texts are available in most countries to reduce the amount of effort needed when drawing up tender specifications. They are arranged by trade and constructed as modules: the descriptive text for a required service is compiled in sequence using a system containing several steps, with information on building type, realization type, realization quality, structural component, type of material, quality of material and dimensions, and possibly on the conditions of realization as well. There is however no guarantee of the technical correctness of the texts, as in principle it is possible to arrive at combinations that would not make sense.

compilation of the standard text collections and the hierarchical structure imposed facilitate the exchange of data between the compiler and the recipient of the tender specification, because of good IT compatibility. All that remains for planners is to check the accuracy of the content.

It is possible to ensure that a description is complete to the greatest possible extent on the basis of the prescribed system using standard texts. But it should be noted that appropriate pattern texts are not available for all special solutions.

Building product manufacturers in particular usually offer appropriate pattern texts, which can be taken over into the tender specification very easily. Planners should remember that manufacturers do not see this as an altruistic service, but use drawing up a tender specification as a way of placing their own product. They are also keen to establish unique selling points to exclude rival products from the competition. So information can be supplied with the tender specifications containing production-related information about the thickness of layers or alloys for a particular product that are unique to one manufacturer, but irrelevant in terms of the product's suitability for use and durability. This often sets the hurdle for finding a possibly more reasonably priced alternative unreasonably high for the tendering firm.

In any case, planners are advised to be careful when adopting a manufacturer's product descriptions. If they are not sure what a particular formulation implies, they should consult the manufacturer or more neutral institutions.

Unique selling points that are irrelevant or unimportant			
Non-measurable descriptions	Concrete information, e.g. responsible inspection authorities	Product- and company-specific descriptions	Production-dependent properties
" ...far exceeding DIN requirements" - "...above average" - "excellent"	"General appraisal certificate from..., No..."	"...elastomer bitumen wielding line, PP fabric on top ... grooved VARIO and film on bottom."	"Manufactured in an extrusion coating process."

Fig.42:
Biased manufacturer's description

Freely formulated texts

Formulating the text freely requires a high level of expert knowledge of the work to be described. This approach is also used to distinguish important from less important information. A range of manufacturers, the appropriate associations or other competent partners should be consulted in order to obtain the information relevant to particular requirements. Planners must apply requirements arising from standards and directives responsibly. The same applies to checking for completeness. To ensure this, the description should cover the following points:

- Description of the work required
- Description of the type of work required
- Spatial frame of reference for the work required (information about the part of the building, but also its location in the building, if this is not clear from the numbering)
- Information about quality (material, surfaces etc.)
- Information about dimensions falling outside the reference unit

\\Hint:
A tender specification can be based on a reference object without excluding alternatives (e.g. "door handle stainless steel 1076 brand FSB or of equal standard"). Equality of standard can be checked by requesting data sheets or samples accompanying an alternative suggestion in the bid.

Systematization

The following scheme systematizes the textual description of work, and can be applied to any subsection of the work in this form. Modular standard texts are based on similar patterns.

_ Building method, building type (production by fitting building materials and components together)
_ Structural element (part of the building forming a room or system)
_ Building materials (required or desired building materials)
_ Dimension 1 (element dimensions, such as the thickness of a wall)
_ Dimension 2 (general dimensions, such as the installation height for a particular piece of work)

Supplementary information

It would also be possible to place other information about the purpose of the work on hints on invoicing or realization techniques in the descriptive text. For example, if a building is to be in reinforced concrete, the building method is directly influenced by the requirement to use prefabricated elements. Equally, legal requirements can form part of a tender specification, for example if material removed is to become the contractor's property when the work in completed.

References

It is customary when using texts to describe work required to make reference to other documents, such as statical calculations, plans, samples or reports. For example, a tender specification text could contain the formulation *"reinforcement as per reinforcement plan"*.

Reference to a drawing is particularly useful in the case of spatially complex situations or complex building sections. For example, if an item describes the construction of stairs with a banister, a drawing will help to identify individual elements mentioned in the text and to understand

\\Example:

Building method:	masonry according to the standard xxxx
Structural element:	for the interior wall in section EG XX/YY
Building material:	with calcareous sandstone blocks yyy standard
Dimension of element:	with a thickness of 17.5 cm
General dimension:	built to a height of 3.00 m.

This subservice can be further explained by complementing the above description with information on quality relating to the structural element (e.g. finished as exposed masonry on both sides) or the building material (e.g. salt-water-resistant finish) or on the building method (e.g. build with prefabricated wall elements).

unambiguously the way they fit together. This means that the bidding firm can check the completeness of the service description, and it is easier to estimate the assembly time required.

Reference to reports makes sense in the case of special sound insulation requirements, for example. In this case the relevant requirements laid down in the report do not have to be described in the individual items, but are defined in an appropriate reference: *"Higher than usual demands will be made on the completed building element according to the appended sound insulation report. These requirements must be met, and considered in the bid price."* In this way, planners avoid possible sources of error when transferring individual requirements into the tender specification.

Item type (ITy)

Different types of works positions can be used in a tender specification. Planners identify a particular item appropriately in the "Item type (ITy) column".

Distinguishing between different item types makes it possible for planners inviting tenders to test the market with a view to optional and alternative services. For example, not all decisions will necessarily have become final at the time the work is put out to tender (e.g. the choice of floor coverings), or the requirements for a particular service will not have been fixed (e.g. installing drainage). Subsequent changes can then be addressed through contingent items. If these works are not ordered until after the contract has been completed, the bidder's prices given as per tender specification are binding.

Standard item

Standard items are always realized. Bidders will provide a unit price and a total price for these in the tender specification.

\\Hint:

If identical descriptions occur in different items, there are two ways of avoiding unnecessary repetitions. Planners can describe the work in full in one item and then refer back to this in subsequent items (e.g. plasterboard stud wall, finish as in previous item, but with double boarding). If identical descriptions apply to a number of items, planners can sum them up and identify them in their preliminary notes (e.g. finish plasterboard stud wall as in preliminary note type A). Preliminary remarks always relate to particular pieces of work and are used exclusively in the context of tender specifications listing the works in full. Fundamentally they are the same as texts with lists of works and should therefore match them in terms of content. There is no conflict with general or special contract conditions.

No.	Text	Item	Quantity	Unit	UP	TP
01.02.02.0001	...Textile floor covering...	BI	30	m²		
01.02.02.0001a	...Parkett...	AI	30	m²		

Fig.43:
Example of alternative items

Contingency items

> 

Contingency items make it possible for planners to test the market in relation to services they may wish to use additionally.

If it is not certain that an item of this type will be realized, it is not provided with a total price in the tender specification and thus not in the final bid price either. If the client orders a contingency item after the contract is concluded, its price will be calculated on the basis of the unit price entered by the bidder. If they are not commissioned, contingency items are omitted without claim to remuneration.

Basic and alternative items

Basic items are items that are fixed for realization. Enquiries can also be made about alternative items. Thus, a basic item is considered to be a component of this service to be carried out, and must be provided with a unit and a total price.

Alternative items can replace basic items if the service described is to be realized in an alternative way. As for a contingency item, an alternative item must also be identified appropriately and provided with a unit price only by the bidder. > Fig. 43 Planners can use such items to make the

\\Hint:
Contingency items should be used only for subsidiary works that are not essential to the overall success of the building project. It would make sense to use contingency items if certain items have not been fixed before building starts because of insufficient information about the soil conditions on the building site.

\\Tip:
Contingency items are often not checked carefully enough when examining the bid, as they are not included in the bid total. This can lead to accepting inflated unit prices that have to be kept to when ordering the realization of contingency items.

No.	Text	Item	Quantity	Unit	UP	TP
01.02.02.0001	...Excavation soil class 3-5...		1000	m^2		
01.02.02.0001a	...Excavation soil class 6...	SI	200	m^2		

Fig.44:
Supplementary item

best financial decision about a service. The basic item is always realized unless the client expressly orders the use of the alternative item.

Supplementary item

Supplementary items are a different item type. They identify potential impediments or additional work needed in relation to a standard item. Bidders provide a unit price and a total price for supplementary items in the tender specification. The corresponding standard item then covers something like a basic finish for the item, and the supplementary item describes a higher standard or a special installation situation. The price for a supplementary item is calculated from the difference between the price for a higher standard and the basic finish.

An example shows the possibilities arising from using a supplementary item. Figure 45 presents two variants for describing the same work for applying exterior rendering.

The first variant shows the rendering in two standard items, separated for the high and the low building. The other provides a tender specification with a standard item for the whole building project and a

\\Hint:
An alternative item offers the possibility for contractors to propose their own solution for implementing a basic item (e.g. "Masonry to be produced according to previous item, but realized according to the bidder's choice.") A description of the proposed finish must be appended.

\\Example:
The example in Figure 44 clarifies the principle behind a supplementary item. The work described in item 01.02.02.0001 covers excavating 1000 m^3 of soil in classes 3–5. The appropriate supplementary item enquires about the price if the problem arises from "excavating class 6 soil" to the extent of 200 m^3. Thus the price contains only the proportion (supplementary price) for dealing with the problem, and does not represent a price in its own right for excavating class 6 soil. Hence the areas (200 m^2) are already included in the standard item.

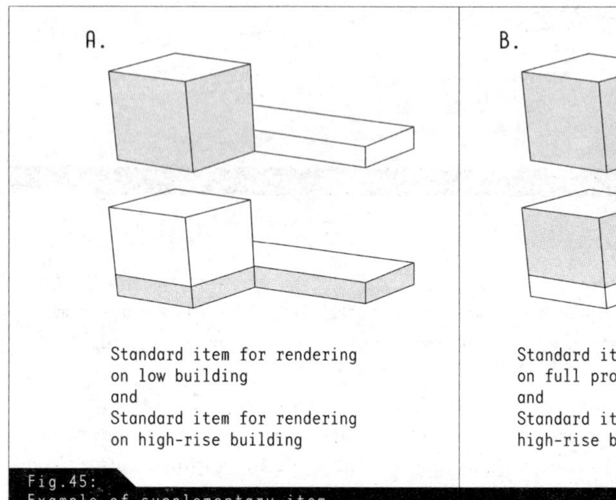

Fig.45:
Example of supplementary item

supplement for the higher areas, for which scaffolding costs may have to be quoted. The second variant has the advantage that a bid will not be made for the ground floor of the high-rise building at a higher price because of additional scaffolding costs.

Quantity and units of quantity

Quantities can be expressed in various units. But the unit of quantity (UQ) should always relate meaningfully to a particular subservice. Thus, it is possible to calculate reinforcement in cubic metres (m^3), although this is a disadvantage in terms of the form reinforcement bars will take and the industry standard of calculating in tones to determine prices. Sensible units are tones, or where applicable square metres (m^2) for steel mats and metres for steel bars.

\\Tip:
Supplementary items often differ only very slightly from the corresponding standard positions. In such cases the descriptive text can be reduced to the essential changes, there is no need to repeat textual elements that remain unchanged. It is sufficient to write "supplement to item xxx for double boarding."

\\Hint:
The general and special requirements contain guidelines for the use of units of quantity. For example, wall areas are calculated by area (m^2) for rendering work and soffits by length (m).

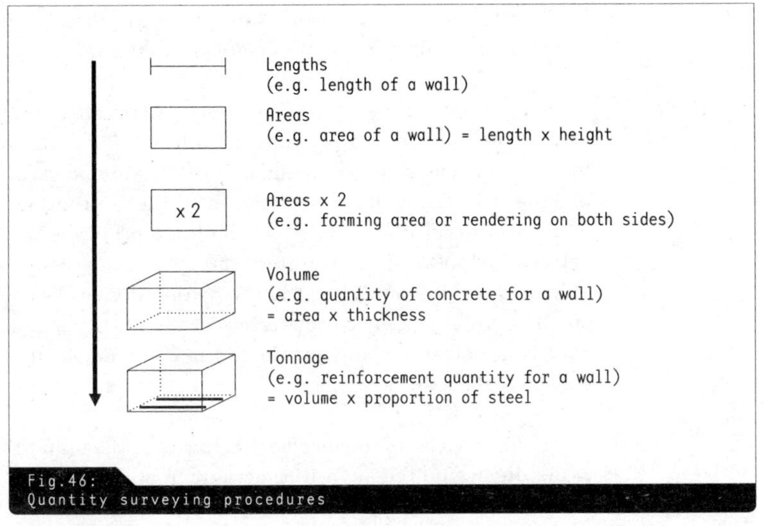

Fig.46:
Quantity surveying procedures

The unit of quantity fixed for an item forms the basis for quantity surveying and is the reference value for bidders when fixing prices for subsections of the work.

Quantities relate to the quantities required for completing a piece of work on the scale required. The quantity is given as a preliminary estimate representing the quantities required for completing the work according to plan. They are derived directly from the working plans, or in the case of refurbishment work also by on-site estimates, and built into the tender specification by the planner inviting tenders. It is fundamentally advisable to record the quantities for a work item within a particular frame of reference (total building project, part of building, floor etc.) and to colour code the corresponding structural elements in the plan appropriately to avoid redundancies. Quantities can also be worked out by using appropriate construction software (e.g. CAD). The degree of automatic quantity calculation extends from simply working out area to a complete component-oriented list of quantities using a 3D model.

Conscientious quantity surveying also makes it easier to account for building works at a later stage.

Unit price (UP)

The unit price (UP) is calculated by unit quantity by the bidder on the basis of the description of the subservice and built into the tender specification. As a rule the unit price is fixed and usually forms the basis for later invoicing for the work. Unit prices are changed only if quantities vary considerably or content deviates from the work as described.

Total price (TP)

The total price is arrived at primarily from the product of unit price and preliminary estimate (planned quantity). Bidders should work it out for all items basic and supplementary items intended for realization and include it in the tender specification in the appropriate place. The sum of all the intermediate totals for a building project is the net final tender price. Adding VAT at the statutory rate gives the gross price the bidder is offering for the commission to provide the services described in the various lists. Total prices are not stated for alternative and contingent items, as it is not clear at the point the bid is being made whether these items will be realized or not.

Subservices are invoiced on the basis of the total prices for the individual items and the quantities actually realized.

IN CONCLUSION

Tendering is not usually one of the activities planners find themselves looking forward to with particular glee. This is understandable, as the charms of a beautiful design, a magnificent view and even carefully planned details seem incomparably greater. The large proportion of text alone often detracts from the allure of the tendering process.

But planners who take the trouble to tender carefully for the ideas in a design will gain some very sound insights into their own planning and the necessary sequences of events needed to realize it.

It is only the invitation to tender that will ensure a high standard of planning is also reflected in outstanding realization.

So this volume is intended to spur planners on to formulate their own invitations to tender comprehensibly, and to structure them meaningfully. If the contractors understand the invitation to tender, the planners have not let themselves down.

APPENDIX

LITERATURE

Bert Bielefeld, Lars-Philip Rusch: *Building projects in China*, Birkhäuser Verlag, Basel 2006

Bert Bielefeld, Falk Würfele: *Building projects in the European Union*, Birkhäuser Verlag, Basel 2005

Institution of Civil Engineers, Association of Consulting Engineers and Civil Engineering Contractors Association: *Tendering for Civil Engineering Contracts*, Thomas Telford Ltd, 2000

CIRIA: *The Environmental Handbooks for Building and Civil Engineering: Vol 1. Design and Specification*, Thomas Telford Ltd, 1994

SAMPLE INTERNATIONAL CONTRACTS

FIDIC	Fédération internationale des ingénieur communauté
NEC	New Engineering contract

ADDITIONAL SOURCES OF INFORMATION

ISO	International Organization for Standardization (http://www.iso.org)
CEN	European Commitee for Standardization (http://www.cen.eu)

In addition to the above-mentioned sources, there are a number of national and international associations and institutions that offer leaflets, examples of additional technical contractual terms, as well as sample tender texts for certain items of work. Sample tender texts that can be used by all types of contractors can be found on the following Internet sites:

Internet sites

http://publications.europa.eu

http://www.neccontract.co.uk

http://www.fidic.org

Tendering portal

http://ted.europa.eu

PICTURE CREDITS

Figure 6 left:	aboutpixel.de
Figure 6 centre right:	PixelQuelle.de
Figure 7:	PixelQuelle.de
Figure 8:	PixelQuelle.de
Figure 10 centre left:	aboutpixel.de
Figure 10 right:	aboutpixel.de
All other figures:	The authors

P11

绪论

从设计到施工

设计师——包括建筑师、土木工程师和专业工程师——最终都要在设计递交和审批过程结束的时候与其他人打交道。在这个过程中，前面的阶段主要集中在与业主和政府部门的交流，但是现在设计师把他们的注意力转向负责建设这个项目的单位：工匠、建筑承包商、专业公司。

招标内容

所有实施项目所需要的信息都会作为招标文件的一部分，以对要做的工作和所需要服务的文字描述或者图纸的形式提供给建筑公司。招标文件必须包括参与报价的公司所提供必要的服务、提出合同报价以及适当的工作计划所需要的全部内容。

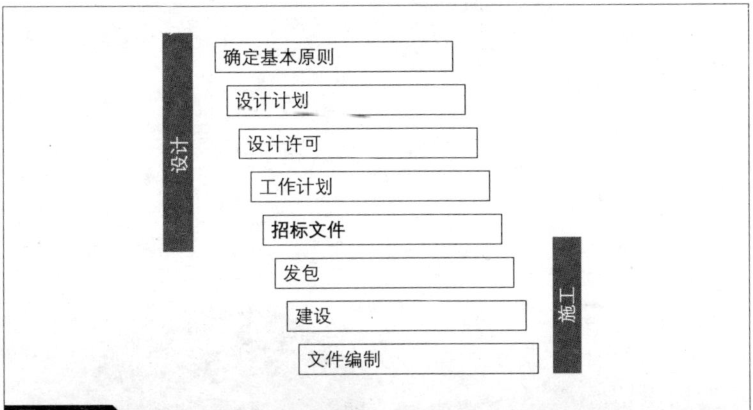

图 1：
设计阶段

提示：

标书是业主与建筑承包商签署的合同的一个组成部分。在招标过程中，设计师会因为错误和疏忽而招来诟病。

招标过程的成果

招标的目的是吸引尽可能多的合理报价以形成对市场的全面了解。招标邀请书由设计师编撰并且提交给相应的承包商，然后他们会背对背地计算出价格并给出一个报价。然后由设计师进行分析和比较。通过比较，设计师能够掌握当时的价格，保证业主能够把项目委托给价格最合理的单位。

P12

招标要求

招标邀请书囊括了设计阶段提出的所有要求。这些要求基本上都是由业主提出的，但是也可能是和法律或技术问题相关的。它们可以分成以下类型：>图 2

— 成本；
— 期限；
— 功能；
— 指标；
— 质量。

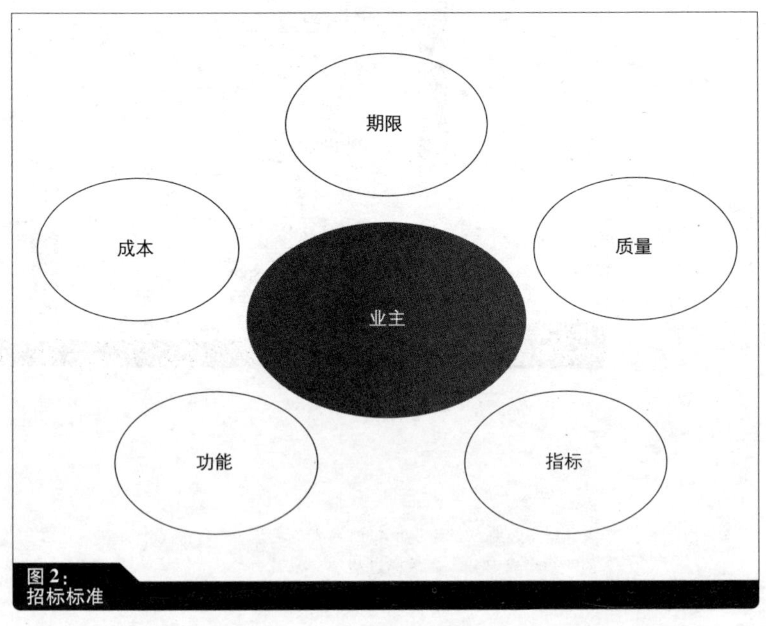

图 2：
招标标准

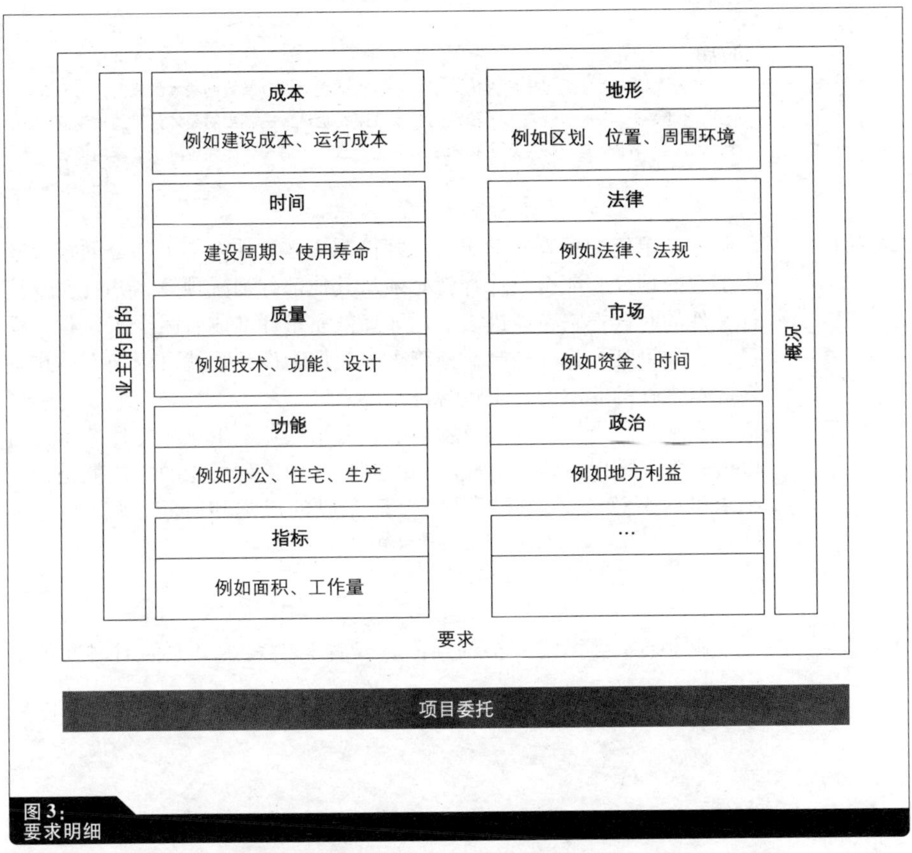

图 3:
要求明细

这个标准用来确定施工进度和其他需要提供的服务。>图 3

P13 成本

成本范围

在大多数情况下，成本都是衡量报价可行与否，甚至是项目本身可行与否的重要标准。设计师必须实现业主投资的利益最大化。设计师通常会有一个预算，必须在此基础上确定建设工作的所有成本。这意味着把不同服务和发包单位的预算整合起来。>见“招标组织”，“确定报价单位”一节

对于业主来说，把开支控制在预算之内往往是整个项目是否成功的关键。首先要做的就是根据设计师按照实际市场价格建议的成本来核对承包商的具体报价。

小贴士：

在首次招标文件中提出的报价对于业主能否建立对设计师控制成本的能力的信心非常重要。如果第一次报价就超出了预定的成本框架，也许业主就会对整体预算的控制感到担忧，也许会对所有其他标准产生影响在前期进行合理的调整，比如说降低装修的标准。

成本控制　通览单笔的预算有助于设计师控制成本。如果某个部分的报价超出了它的预算，那么设计师就必须从其他部分削减预算，并且在发招标文件的时候把它考虑进去，比如降低质量标准或者减少工作量。反过来，如果某个部件控制在预算之内，那么设计师就可以把先前排除在成本之外的业主的要求包含进来。

成本保证　在施工、市场开发以及发包的过程中避免发生意料之外的事情可以最大限度地保证业主对成本的控制。其中一种可能的方法就是让一个承包商来总负责，这样可以保证完成的成本和时间。> 见“招标组织”，“确定报价单位”、“打包发包”等章节

P15

期限

业主通常会设定严格的期限，或者至少表达出他们对时间的要

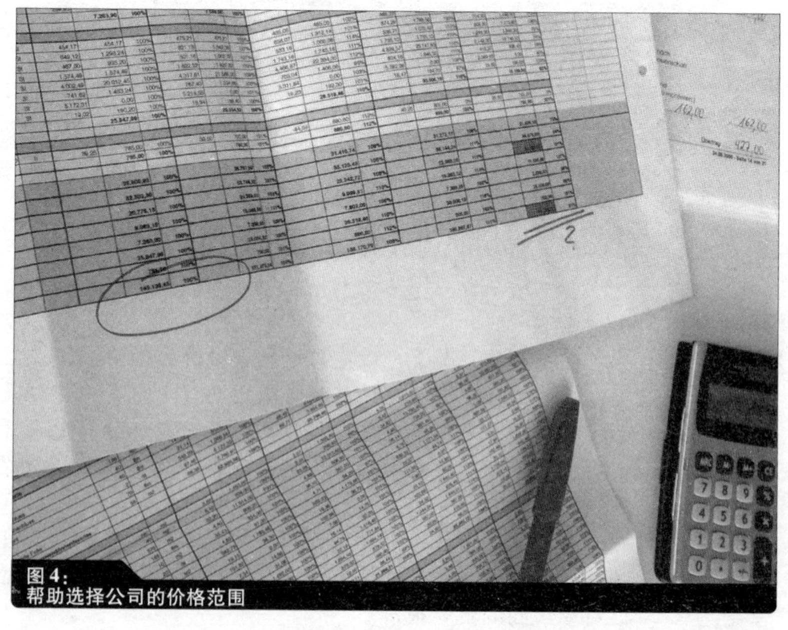

图 4：
帮助选择公司的价格范围

求。期限一旦经过确认，就必须遵照执行。

时间限制

时间限制主要是根据具体建筑的设计用途来确定。例如，完成日期，以及由此而确定的入住日期，对于建设私宅的业主来说都是非常重要的，因为他们要恰当地处理先前租来的房子。学校的翻新项目只能在学校放假的时候进行。在这种情况下，开工和完成的日期是决定性的因素。

作为成本因素的期限

时间要求可能也会对施工过程产生影响，并进而影响成本。加快进度的惟一实际办法就是增加雇员、设备和材料的数量。因此承包商不得不租借设备或者加班加点，在周末甚至晚上还要工作。这会导致报价的提高，因为要把额外的成本含在报价内。

对招标的影响

时间限制也会影响设计师提交标书的方式。丰富而详细的策划需要投入大量的时间，所以设计师必须考虑他们是否能在适当的时间内提交这样的策划。如果他们做不到，他们就会通过限定某些功能性条款，把某些策划工作转给承包商，而不是详细地完成策划。>见“招标组织”，“标书风格”，“功能性标书”等章节

小贴士：

设计师必须保证业主对期限的想法是符合实际的。这会影响到相关公司所提供的服务以及设计师自己的能力表现。施工过程中的一些事情很难改变或者根本就无法改变。这包括得到政府部门的许可、天气以及一些产品的运输时间。

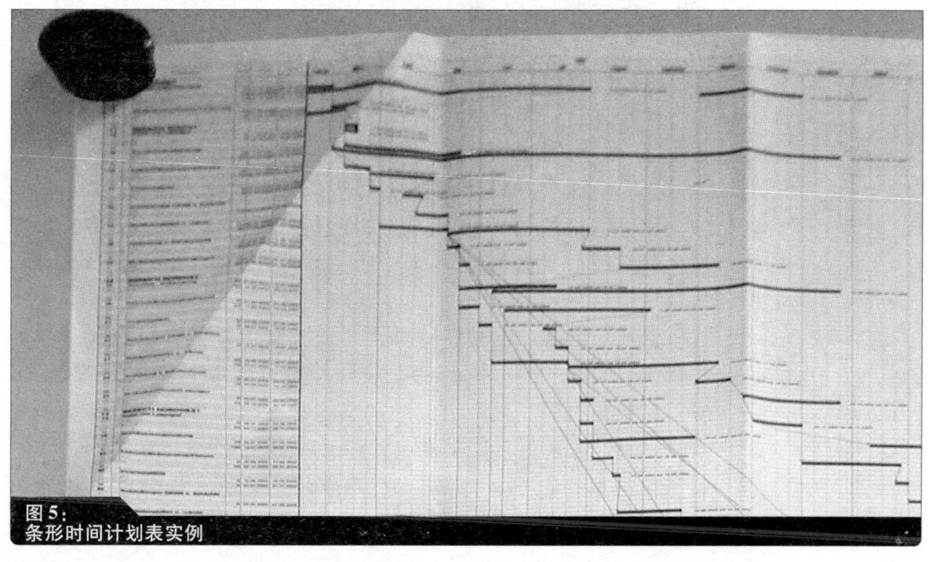

图5：
条形时间计划表实例

图 6：
不同的功能

P16 功能

实施范围

业主的要求决定了实施内容的深度和广度。例如，如果私人业主想要买地盖房，那么他可以建阶梯式住宅、半独立式住宅或者是独立式住宅。因此功能是决定建筑形式的要素之一。也可以根据功能描述来决定建造方法。对于没有特殊要求的仓库来说，通常会根据价格选择一种形式（比如说钢筋混凝土或者是钢结构）。因此，功能和由业主的具体要求所决定的可能的解决方案密切相关。

业主简介

业主对项目的确定性越强，他们就越希望在招标文件的策划中发挥作用。如果这个项目是他们梦想中的家，那么他们就会非常希望参与策划过程中的每一个细节。因此招标文件必须非常详细，这样才能完全实现业主的想法。>见“招标组织”、“标书风格”、“详细的标书”等章节

如果建成的建筑是营利性项目，那么，业主主要关心的是以最小的投入换来最大的产出。他们会在一开始尽可能地减少他们的想法，只有在会有更高的利润或者更大的销路的时候才会考虑提高技术和美观的要求。如果业主只是想简单地把一些东西放到一个方盒子里（例如一个仓库或者一个工业生产车间），他们就会非常实际地从功能角度出发，而不想用太多的细节来麻烦自己。

图 7：
有不同的方法来满足建筑的要求

P18 服务内容

最低限度 服务内容的范围取决于业主的要求。例如，对于写字楼来说，业主会提出需要多少个工作站，以及需要什么样的空间来服务于既定的功能（门厅、会议室、服务区等等）。要求越精确，能够确定的项目最低限度也越精确。

合理化 如果项目的内容与既定的成本框架不匹配，设计师可以通过合理化来降低成本（例如，尽可能大量采用相同的构件，选择同一个服务提供商）：立面设计要符合生产商制造的立面板材的规格，这样可以避免出现大量的切割或者特殊规格的定制。

容易受影响的因素 和最低标准决定的范围一样，还有一些变量也会对建筑的质量产生影响。例如，设计师可以减少窗户的面积，因为窗户的价格比封闭的立面要贵得多，代价是降低使用者的舒适度，或者通过牺牲后期的灵活性来减少工作站的数量。

P19 质量

功能也会影响预期的质量。这里我们指的是技术和美观的标准。>图 9 技术要求包括建筑的法律规定（例如与人员集散或者与消防有关的法规）、健康要求（例如通风或者卫生）；美观要求与视觉影响、

图 8:
功能和质量的关系

作为一个整体的建筑形式和特征，甚至门把手之类的单个细部有关）。

质量标准

对于大多数建筑服务来说是有着明确的最低标准的，以保证采用适当的材料和专业的施工。业主会对自己的财产有一些高于最低质量标准的要求。一旦策划的装修背离了标准的质量，设计师必须在他们的服务描述中清楚地表达出来，并且对这种装修和预期的效果进行描述。

质量	
技术	美观
结构 建筑科学 静力学 法律 …	形式 视觉 触觉 …

图 9:
技术和美观标准划分

图 10：
任何东西都可以包括在招标文件之内

如果与装修质量有关的要求超出了普通标准，成本就会跟着增加。例如，档次比较高的墙面装修的抹灰工作量和成本都会比在连接部位简单找平和装修要高得多。

长期策划

与质量有关的问题需要予以长期的重视。例如，安装一个虽然价格较高但是容量较高的采暖系统，可以很快通过使用过程中的低能耗来弥补它较高的成本。

P20

标书条目

建筑服务和建设成果

建设过程包括大量结构构件的选择和协调。在这里，设计师可以在他们的计划中列出很多预制构件（比如门和门框），但是也可以单列每个部件（比如手工打造的门配件）。建筑可以细分到每一个螺丝钉的位置和它的头型。像建筑服务商发出的招标文件既和策划过程中的各个部分有关，也和整个施工过程有关。它们概括了所有需要提供的服务。根据建设项目的规模和标书特点的不同，招标文件的内容也各不相同。>见“招标组织”，“标书风格”等章节例如，如果招标项目是一个从草图到建成的完整的建设项目，那么标书应该包括从基础开挖到清理施工现场和提交钥匙的所有服务内容。招标文件也可以只是要换一个窗户。

策划和其他服务

招标文件里还可以包括补充的策划和服务，以及标准的建设工作或产品。这样它可以请人做一个隔声报告或者组织一个一流的聚会之类的专业报价。

提示：

招标、收集报价以及由业主委托项目之后的阶段叫做发包程序。

可以在《Building Projects in the European Union》一书中找到有关投标发包的信息，Bert Bielefeld，Falk Würfele，Birkhäuser Verlag，2005。

P23

招标组织

组织的基本原则

项目的规模以及招标文件的多样性和复杂性意味着把建设过程分成几个重要的阶段是一种非常明智的做法。为了达到这个目的，设计师必须熟悉施工过程中的各种事情和每件事情的相互依赖性，这样才能按照正确的时间顺序把它们排列出来。

P23

列出招标时间表

期限和标书

期限是招标过程中的一个重要因素。设计师必须了解多长时间能把项目完成。然后根据施工期限列出设计师和报价公司的工作时间表，同时必须考虑相应的专业公司前期策划和发包程序所需要的时间。

P24

参与人员的时间投入

设计师投入的时间

设计师需要足够的时间来编写招标文件。>图 11 一旦他们按照业主的意愿和要求列出了一个清单，他们就必须花时间组织招标邀文件，并且考虑如何在标书中说清楚这些要求的问题。设计师必须提出质量要求，以明确需要提供的服务范围。生产商、专家组或者是其他的承包商必须解决所有出现的问题。通常必须提供概况的专业介绍和图纸，以确保所提供的服务能够充分满足要求。如果会涉及到比较困难的安装或者复杂的施工，那么对于设计师来说比较明智的做法就是通过书面或口头咨询制造商来完成这个任务。

一旦设计师列出了所需要的服务，他们就必须在和业主协商之后整理出一个报价者的名单，也就是说，所有被邀请参加投标的公司的名单。招标文件必须复印后分发给相关公司，保证他们有足够的时间来完成任务。

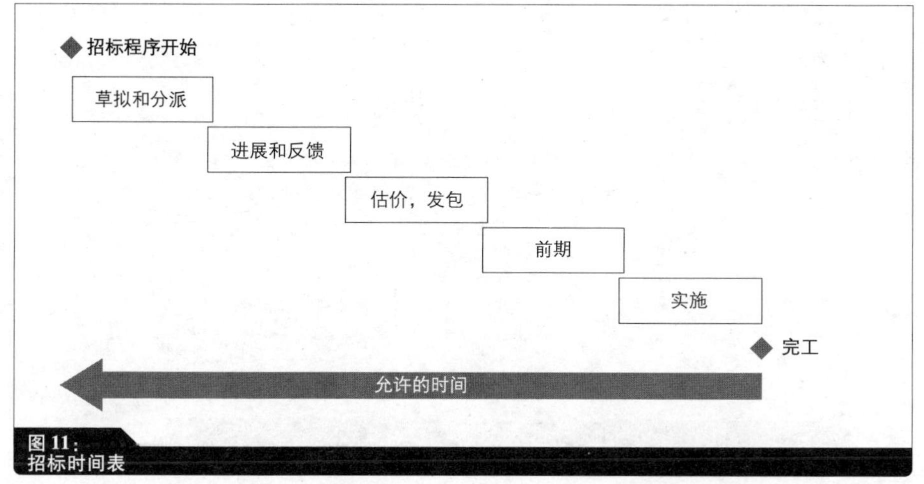

图 11：
招标时间表

报价过程所需要的时间

收到任务书后提交报价的施工单位必须熟悉新的建设项目，因此需要花一段时间来研究招标文件。在某些情况下，在最终计算价格之前还需要对标书的特征和范围进行额外的策划工作；需要分别考虑生产商或者其他的公司以及他们的内部询价。价格计算必须把工费、材料、设备和室外作业考虑在内。除了这些与项目直接相关的因素之外，还需要在报价中包括企业一般管理费用和可能的利润。根据项目的复杂程度和规模的不同，准备报价的时间可以从一天到几个星期不等。如果招标文件还包括了策划服务或者技术测试的话，需要的时间就会相应长一点。

前期行业策划阶段

从项目委托到施工现场服务的实际交付（开始施工）之间的时间段是专业公司的前期行业策划阶段。> 见图 12 在这段时间里，受委托的公司可以编写工作计划或者制作样品，并且让他们得到设计师的认可。结构构件通常会由相关单位提前在工厂里加工或组装。预制构件的尺寸需要根据现场决定，这就又增加了一个影响完成时间的因素，例如只有砌块墙的洞口完成之后才能确定预制窗户的尺寸。

施工所需要的时间

在排标书时间表的时候，如果要满足完工时间的要求的话，了解完成某一项工作需要多长时间是非常重要的。> 见图 13 这些单项工作所需要的时间几乎没有什么压缩的余地。工人的数量、工作的时间和机器的使用会影响所需要的时间。工作速度的加快有一些自然的限制，比如说某种建筑材料干燥或者硬化的时间（例如抹灰）。施工现场的空间也是非常宝贵的，所以增加劳动力也许会导致工人在工作的时候相互干扰。

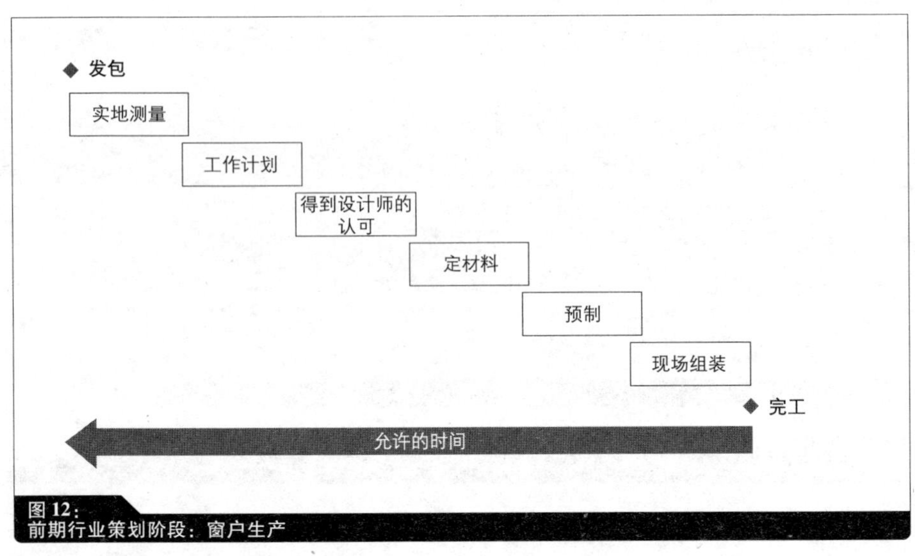

图 12：
前期行业策划阶段：窗户生产

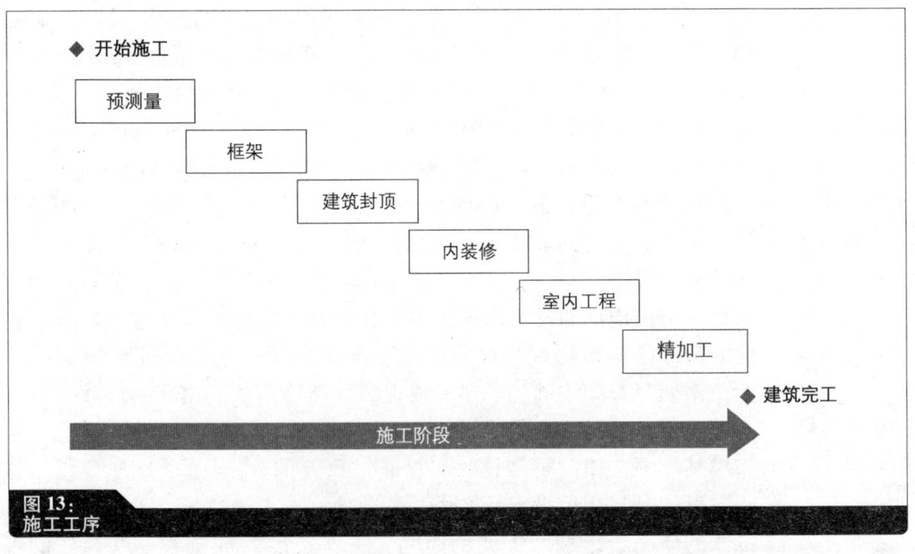

图 13：
施工工序

P26

招标程序和招标类型

时间程序

招标文件的编写顺序一般是根据现场的工作顺序组织的。>见图13首先是实际建设项目准备措施的报价，然后是框架施工、外装修、室内装修的报价，再接着是精加工的报价。有时候前期的准备工作会跟实地的程序相背离。例如，立面的施工可能需要长达数月的准备期，这是在编写标书的时候必须考虑进去的。

建设进程中的招标文件

这里所说的是与施工过程中发出的标书有关的程序。>图14对于这种标书来说，所有的服务都是按顺序排列的。例如，室内工程是在框架工程快要完工的时候策划和招标的。与综合性标书相比>见“招标组织”，“确定报价单位”，“打包发包”等章节这种方法的好处是能够对预料之外的成本作出适当的反应。>见“绪论”，“招标要求”，“成本”等章节它还可以把在已经完成的施工阶段中出现的问题包括在内。比如说，如果由于某种原因必须改变顶棚的厚度，那么可以通过装修完成面的楼板高度来加以补偿。然而，这样只有在最后的招标完成之后才能确定成本，因为之前很难预测市场波动和其他可能发生的事情对所提供的服务产生的影响。

综合性标书

委托给一个主承包商能够更好地控制成本。>见“招标组织”，“确定报价单位”，“打包发包”等章节在这里，所有的服务都必须充分明确，并且和招标文件一起发给承包商。相应的策划阶段也会时间长一点。之前和施工同时进行的策划工作现在必须在开挖之前就完成。>见图15如果没有足够的时间来做详细的招标文件，那么设计师把精力集中在

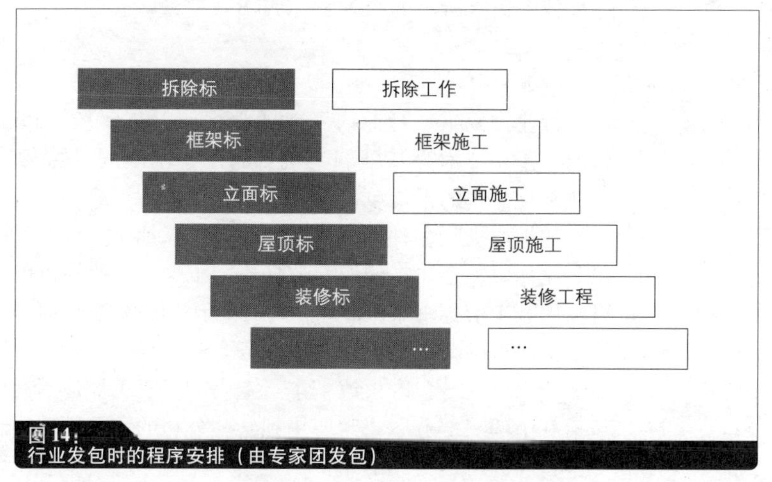

图14：
行业发包时的程序安排（由专家团发包）

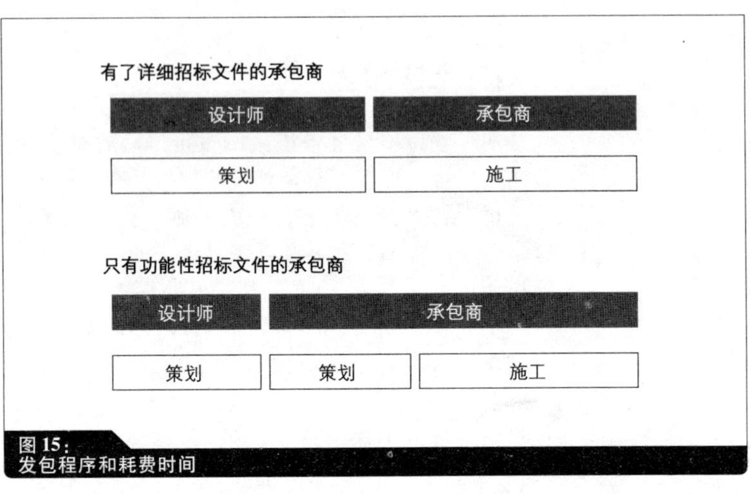

图 15：
发包程序和耗费时间

对于业主来说比较重要的那些要求上，并且单把它们详细列出来 >见“招标组织”，“标书风格”，“详细的标书” 等章节而其他的只是功能性的说明 >见“招标组织”，“标书风格”，“功能性标书” 等章节。有时候时间太紧，所以只能考虑功能性的标书模板，根本没有什么细节。

P28
报价单位

确定报价单位

报价单位确定了特定承包商的服务范围。报价单位可根据大小分为单独的行业（行业分包）、分块承包和总体承包，也可以在一个总体承包中包含所有的服务。>见图 16

P28
行业（专业分包）

行业分包

行业再分工（行业或者专业分包）是以传统上由个人或公司交付的工艺和技术为基础的（例如石工、木匠或者抹灰等技术行业）。这通常是最小的报价单位。>见图 17

如果要提供多种不同的服务，也可以把一个行业分解成更小的单位。比如说，金属工的招标文件囊括了这个行业提供的所有服务。也可以用若干份标书来限定单个工序，比如说金属立面施工、钢楼梯和扶手。最小的报价单位是单项服务。>见图 18

所以行业分包既可以把本行业内的所有工作包括在内，也可以只包括其中的一部分。当需要用到某些公司的专业领域时，它有利于把一个行业分解成更小的单位。

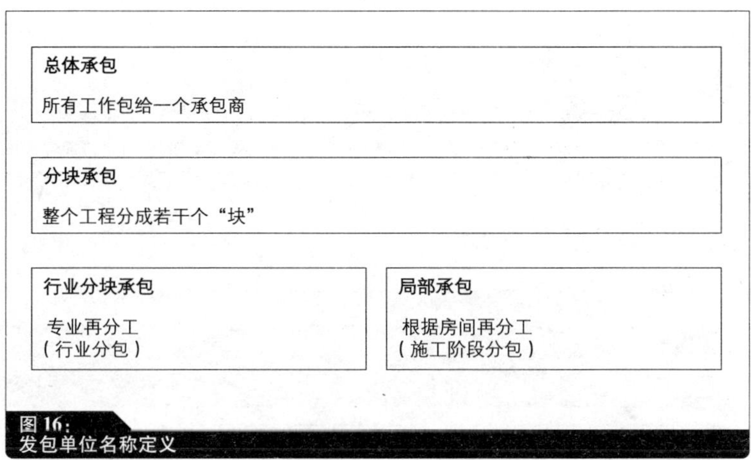

图 16：
发包单位名称定义

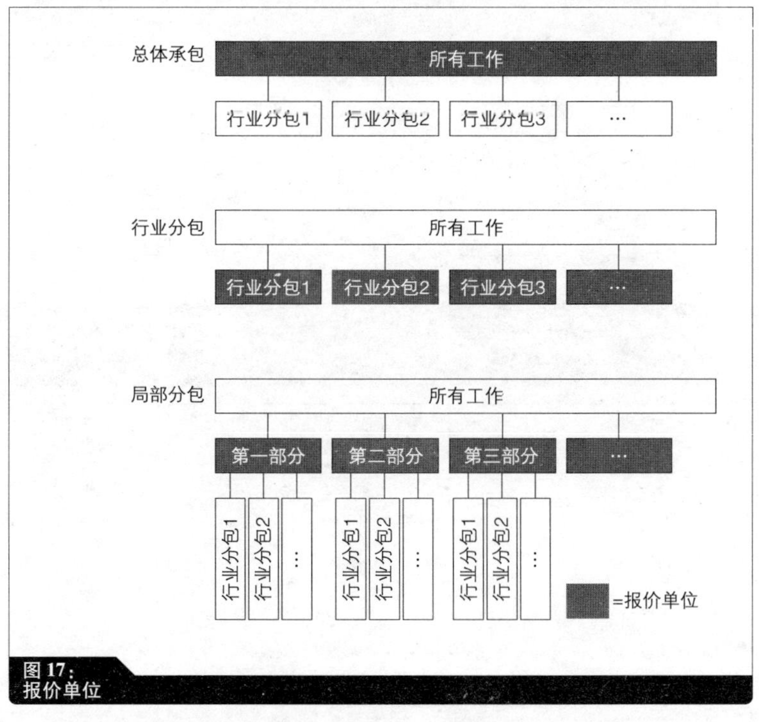

图 17：
报价单位

缺点是当委托公司较多的时候，它需要投入更多的时间和精力来协调，配合性会比较差（例如去现场或者以更合理的价格批量运输）。

提示：

行业这个词通常也用在结构工程、媒体策划或者标志设计之类不那么传统的工种上。尽管它们不是传统的行业，但是这里所指的主要特征就是构成一个单位的工作。

实例：

每天生产和组装楼梯的施工单位能够提供比虽然涵盖该行业各个方面但是主要负责立面金属工艺的单位更加专业和更加便宜的服务。

准备措施

— 建设场地准备
— 拆除工作
— 清理场地
— 开挖
— 场地准备
— ……

框架

— 开挖
— 砌筑工程
— 混凝土施工
— 钢结构施工
— 密封工作
— 木工
— 脚手架工程
— ……

建筑维护结构

— 木工
— 钢结构施工
— 密封工作
— 屋顶工程
— 管道工程
— 保温工程
— 抹灰
— 立面工程
— 金属工程
— 玻璃工程
— 油漆
— 脚手架工程
— ……

装修

— 抹灰
— 找平
— 地板饰面
— 混凝土砌块
— 天然石材
— 瓷砖
— 镶木地板
— 金属制品
— 干施工
— 细木匠活
— 油漆
— 脚手架工程
— ……

室内工程

— 采暖设备安装
— 通风设备安装
— 卫生设备安装
— 电气设备安装
— 电梯
— 媒体技术
— ……

精加工

— 建筑清扫
— 装锁
— 室外空间
— 场地清理
— ……

图 18：
行业再分工

建筑行业

有时候把几个行业捆在一起会比较明智。由一家公司承担与屋顶相关的所有工作比较合理，而且可以避免多个公司的协调工作。因此，木工活（搭建屋架），屋面工作（从保温层到瓦）和一些金属配件（排水沟、起保护作用的铅框）的工作可以由同一家公司来完成。许多公司为了适应业主的要求会设立一个联系人，并且打出能够提供全方位服务的广告。这里需要说明的是有些看上去很大的公司只是简单地“买进”服务，所以并不能给出特别合理的价格。那样业主只是通过给所委托的公司增加组织内部再分工的费用来买一个方便。

>

>见“招标组织”，“确定报价单位”，“局部分包招标文件”等章节

P31

局部分包招标文件

局部分包

局部分包是另一个报价单位。在这里，服务不是按照行业分的，而是按照不同部分来分的。这些部分主要是根据工作量较大、需要几家公司来完成的要求的决定的。

再分工

在公共项目中，这可能会和想要把尽可能多的公司包括在报价过程的想法一起出现，因为之后会根据公司的平均水平要求服务的范围。

施工阶段

另一种适合采用局部分包的情况是长期的策划工作有可能会被打断的时候。对于大型建设项目来说，施工的时间通常是固定的，因此当其他部分还没有最后完成的时候，某些部分可以提前使用。>见图19

P31

打包发包

主承包商招标

正如上面所提到的那样，把几个行业捆绑在一起，只跟施工方的一个代表签订合同，然后由主承包商负责追踪。>见“招标组织”，“确定报价单位”，“局部分包招标文件”等章节在这里，业主委托一家建筑公司提供完成建筑所需要的所有服务，只签署一份而不是多份合同。

提示：

转包合同是已经跟业主签署了施工合同的公司所采用的。转包合同跟业主没有法律关系。由受委托的公司负责委托、完成、付费和保证。

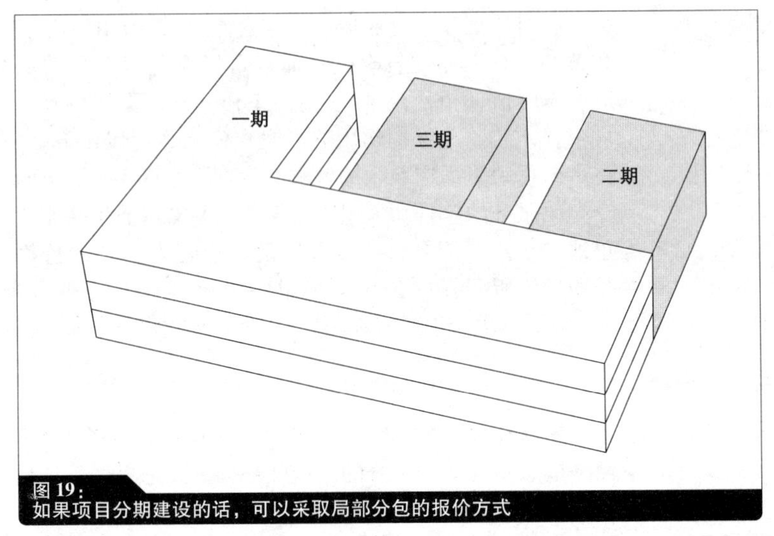

图 19:
如果项目分期建设的话，可以采取局部分包的报价方式

满足时间要求

由一家公司负责控制完成时间会相对容易一些，因为这样的公司可以通过在整体协调中加快其他部分的进度来弥补项目中某些部分造成的延误。把项目委托给几家公司的时候要控制整体时间会难得多，因为其中会有大量的相互依赖性：单个的公司彼此间没有合同上的约束性。

主承包商补充合同

这项工作由主承包商负责，通常这些大量的协调工作会有额外的报酬，作为报价的补充，有时候还会有担保的风险储备金。在实际操作中，几乎没有什么公司能够只用自己的工人完成所有的工作。实际上，他们倾向于外包给其他的公司，而那些公司——如果受到委托的话——就会承担分包商的角色。主承包商的报价中包括了按合同要求时间完成任务的保证金。

P33

标书风格

功能性标书和详细标书之间是有区别的，但是在实践中它们并不是完全分开的。任何详细的招标文件都会包括功能性的元素。例如，即使是对石膏板螺栓墙非常详尽的描述也不会提高石膏板安装的精确度。它假设工人已经有了适当的技术知识而且知道把板材安装到框架上的正确方法。功能性的招标文件可以没有详细的内容，但是应该在这里更加详细地提出要求。设计师向业主提出的关于要求方面的问题越多，实际的功能性标书中详细的要求清单就越多。

P33 功能性标书

功能性标书不会描述如何完成某项工作或者精确的施工过程，而是关注想要得到的结果。即使在最初的报价中没有提到，也是由投标人负责工作的安排并且承担完成想要的结果的责任。除了承担可能的安排错误的责任之外，承包公司还要承担质量检查的风险。

投标人可以根据他们的专业知识和经验来决定如何完成工作。他们可以从自己的资源出发，在合同允许的变动范围内，改变提供服务的范围。

核定报价 核定报价的标准包括价格和既定要求的处理方式。投标人在报价上花了时间和精力，现在设计师必须进行深入的核定。这样，业主或者设计师在接下来的实施过程中就基本上不会再有什么影响。失去控制，尤其是对于详细的标书来说，就意味着失去设计质量。

选择功能性标书 通常会在时间比较紧张的情况下选择功能性标书。>见“绪论”，“招标要求”，“期限”等章节因为它能够明显地缩短委托给主承包商之前的策划过程。功能性标书中业主对施工细节的要求比较少，而承包商的风险也会比较大。另一个选择功能性标书的理由也许非常简单，那就是设计师不知道如何用详细的招标邀请来达到想要的目的。因此，设计师不会就空调厂家的某个部门或者是他们安装的方式进行招标，而是简单地描述出制冷或通风的要求。

P34 详细的标书

详细的标书要求最大限度地把每一个细节都提前考虑在内。设计师不仅要描述想要的结果，还要说明达到这种结果的方法。因此由他们来承担最后的成果没有达到要求、或者是招标文件中的错误和疏落或者是交代不清的责任。额外的工作（需要但是没有包括在最初的招标文件中的工作）会造成成本的增加。

核定报价 核定详细的标书会简单得多，因为选择程序是一定的，只要比较价格就可以了。

选择详细的标书 如果业主想要在施工过程中保持对它的控制，那么详细的标书就是一个明智的选择。这是一个检查施工过程中每一个细节、避免出现不尽人意的意外的办法。

P34 标书深度

从根本上来说，把功能性标书和详细的标书混合在一起是可能的。这为设计师带来了很大的发挥余地。他们可以提交对业主来说非常重要的详细的整体工作计划，并且以同样精确的语言描述出施工的过程。对于那些不需要这么详细的部分，他们可以只是提出要求并且选择能够最好地解决问题的承包商。

详细的标书还是功能性标书?

如果设计师要做详细的标书，那么他们必须有足够的知识。他们要对他们所做出的描述中出现的任何错误负责。因此，比较明智的是在碰到他们不完全了解的东西时，把标书建立在功能的基础上，时刻牢记想要的结果。

完成报价

设计师必须不断地问自己他们所准备的标书够不够完善，或者说他们所提供的信息够不够清楚，是不是没有遗漏。比如说，如果想要立面的定位螺孔是“有序而对称的”，那么他们必须提供表示螺孔模式的图片，以免出现对这个要求的有争议的理解。通常，只有详细的描述才能保证对完成工作的方式的控制。这是非常节省时间的办法，但是并不能用来管理建筑的每一个方面的。设计师必须仔细考虑要详细到什么程度。例如，对于清水混凝土框架的要求要比与地下室框架有关的要求严格得多，因为那一部分在建筑完工之后是看不到的。

P37

招标文件组织

招标文件——不管是功能性的还是详细的——都是由若干元素构成的。>见图 20、图 21 它包括签署建设合同所需要的所有文件。

P37

文本部分

文本部分指的是所有采用文字和图形描述所策划的建设项目顺序和实施方式的内容。招标文件除了这些组成元素之外，还包括下列内容：

提示：

对于非常明确的招标文件来说，不让各个部分彼此冲突是非常重要的；这可以通过确定单个要素的等级来避免。因此，对想要的服务进行精确描述的要求总是高于技术要求的。

— 项目概况；

— 合同条款；

— 技术要求；

— 基地条件的信息；

— 对需要做的工作的描述。

文本部分通常占据了报价的很大一部分。设计师能够从文字中得到计划中看不到的信息。

P37

建设项目概况

概况描述

一份完整的标书包括建设项目概况和发包程序。这些内容会列在封面表或者包括招标文件和其他与发包相关的细节以及主要参与人员名单和对建筑的简短描述的附信之中。

封面表/附信

封面表既是招标文件的介绍，也是内容的概述。它包括建筑公司在投标过程中所要提供的所有信息的综述和申请的条件。封面表应包括下列信息：

— 文件收发人的详情；

— 日期；

— 建设项目的定义；

— 所要提供的工作的位置、性质和范围；

— 与建设项目相关的基本情况（发包形式、时间表等）；

— 明确的招标文件；

— 申请条件；

— 所含文件清单（合同文件）。

建设项目的描述

需要对所要提供的服务位置、性质和范围以及建设项目的定义进行详细的描述，这样才能把它们清楚地界定出来。这些内容要尽可能简明扼要。如果需要项目的更多信息，应该把它们放在建筑描述里面。>见“编写服务说明书”，“编写功能性服务说明书的方法”，“有设计的功能性说明书”等章节

关于发包程序的信息

封面表中要对发包程序进行清楚地描述并且必须受到大部分业主的欢迎。>见“招标组织”，“列出招标时间表”，“参与人员的时间投入”等章节

看现场的日期

确定地点跟可能的看现场的日期有着非常重要的关系。这个日期需要明确在封面表里，给出时间和地点。对于其他没有包括在招标文件中的检查日期来说也是如此。

投标阶段/提交

投标阶段的详情、提交日期和标书装订日期有着同样的重要性。>见“招标组织”，“列出招标时间表”等章节投标阶段决定了递交报价的日

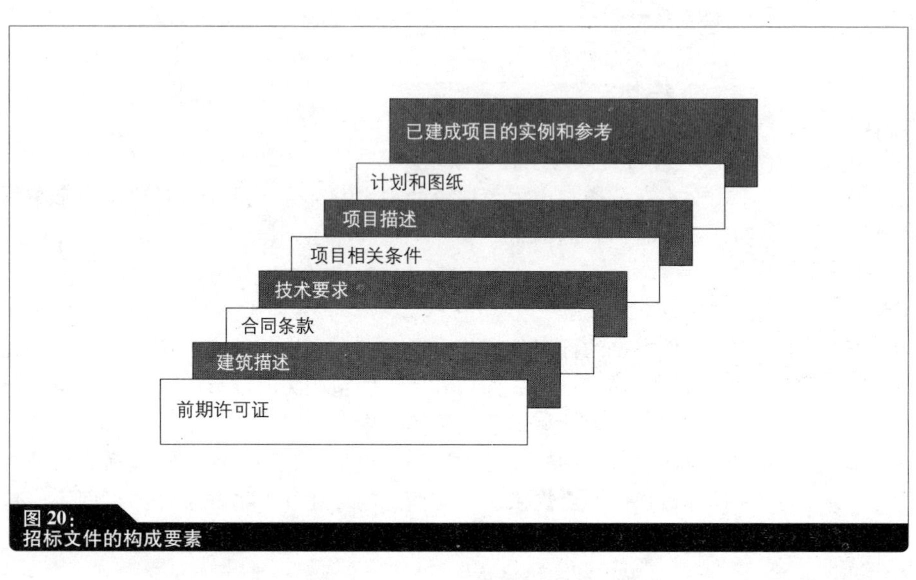

图 20：
招标文件的构成要素

	文本部分	图纸部分	混合部分
具体要素	— 功能描述 — 详细的工作描述 — 发包、谈判和会议记录 — 建筑描述 — 初步意见 — 基地情况 — 报告 — 适用的特殊合同条款 — 适用的特殊技术要求	— 计划 — 草图	— 测试 — 样品 — 参考物品 — 建成实例
标准化要素	— 标准的服务文本 — 一般合同条款 — 特殊合同条款 — 通用技术要求 — “采用的技术原则” — 生产商清单	— 参考图纸 — 关键细部 — 生产商细节	— 参考图纸 — 关键细部 — 生产商细节

图 21：
招标文件构成要素的系统化

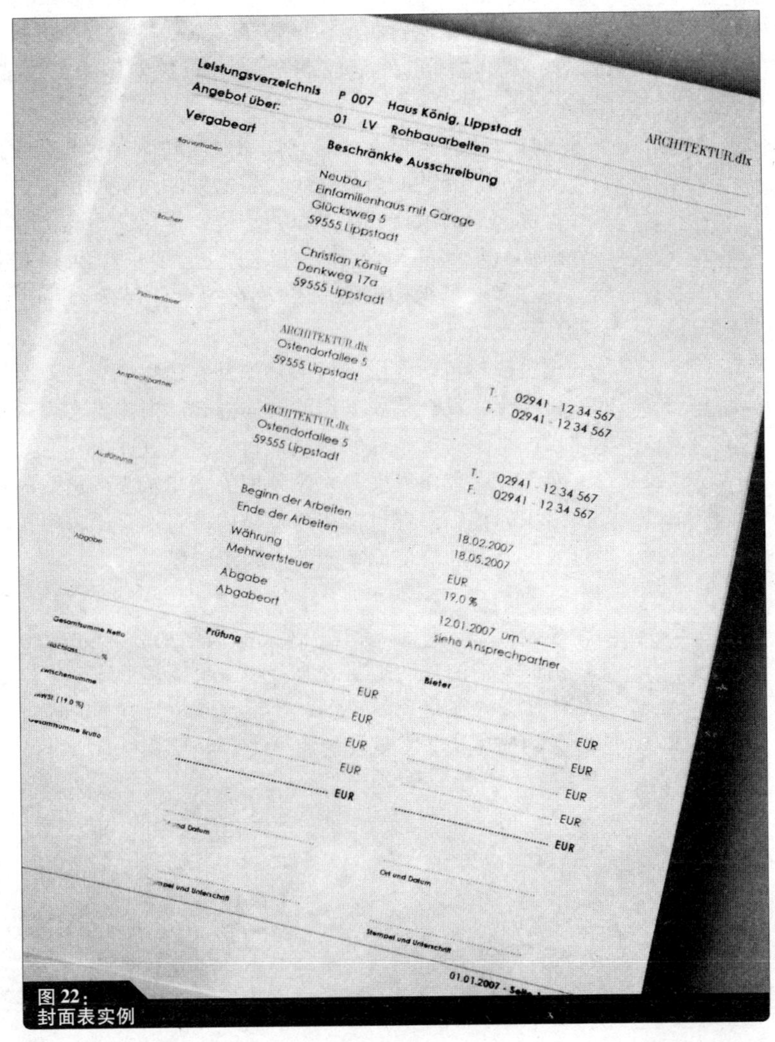
Leistungsverzeichnis P 007 Haus König, Lippstadt
Angebot über: 01 LV Rohbauarbeiten
Vergabeart Beschränkte Ausschreibung
ARCHITEKTUR.dlx

Neubau
Einfamilienhaus mit Garage
Glücksweg 5
59555 Lippstadt

Christian König
Denkweg 17a
59555 Lippstadt

ARCHITEKTUR.dlx
Ostendorfallee 5
59555 Lippstadt
T. 02941 - 12 34 567
F. 02941 - 12 34 567

ARCHITEKTUR.dlx
Ostendorfallee 5
59555 Lippstadt
T. 02941 - 12 34 567
F. 02941 - 12 34 567

Beginn der Arbeiten 18.02.2007
Ende der Arbeiten 18.05.2007
Währung EUR
Mehrwertsteuer 19.0 %
Abgabe 12.01.2007 um
Abgabeort siehe Ansprechpartner

Prüfung
.... EUR
.... EUR
.... EUR
.... EUR
.... EUR

Bieter
.... EUR
.... EUR
.... EUR
.... EUR
.... EUR

Ort und Datum
Stempel und Unterschrift

01.01.2007

图 22：
封面表实例

期。报价人必须在规定日期前提交报价。报价人应该了解招标文件封面表中的基本发包标准（比如说价格）。

约束报价的性质

为了避免误会，封面表应该包括一段说明报价人报价后能够得到的成本的陈述。这段陈述可以这样写：

“我们要求你对这个建设项目进行捆绑报价，我方和业主方不支付费用……”

申请条件/可容许的标准

设计师可以通过附信中的申请条件对可能参加投标的申请人资格加以约束。这些条件可以调节分包商的使用，允许或者排除集体报价。

论证适宜性

设计师可以通过论证适宜性的要求和准入标准来检验报价公司是否具备完成任务的能力。除了已经通知的适宜性检验之外，例如参考已经准确完成的项目，还可以要求他们提供公司经济情况的信息，以确保他们有足够的现金流。习惯上还会要求提供关于建筑公司的资质，它在专业组织中的身份以及它的责任保险，说明它的最小赔偿金额。

建筑描述

建筑描述包括建设项目的其他基本信息。它给公司提供开展工作的项目简介，没有单项工作的具体内容。报价人必须用施工描述和影响建设成本的关键条件的信息来自己完成具体内容。

对于大多数建设项目来说，把关于建筑或者是结构和单个建筑部分的组织的精确信息包括在建筑描述里是一个比较明智的做法，这样报价人可以更加清楚地掌握各个施工阶段。

P41

合同条款

招标文件的目的是要为业主和一个或多个施工单位之间建立合同关系作准备。在这个前提下，控制合同执行方式的规定以及对所需要的服务的描述是非常重要的。

举例：

“项目涉及的是一栋3层的办公楼，总建筑面积约10000m²，为清水钢筋混凝土框架结构。建筑基地位于××街和××街之间，只能从××街和××街交接处的入口进入。精确的基地位置见总平面图。”这个例子说明了建筑描述的内容可以多么简练。它提供了预期的施工方式（清水混凝土框架结构）、工作的性质（钢筋混凝土建筑）、大小（总建筑面积，3层）、功能（办公楼）和基地位置（参见总平面图）。

小贴士：

相关规定通常都非常复杂，因此采用专业机构推荐的合同，或者至少请律师为比较大的项目拟定合同条款，是一个比较明智的办法。

设计师在招标文件中通过一般和特殊的合同条款，为未来的合同作好了准备。一般合同条款可以从完整的合同样本中得到，但是通常来说，特殊合同条款必须根据业主制定的商务条件来设计。

一般合同条款

一般合同条款是以国家或者国际的建设项目管理标准为基础的。它们包括的重要信息有：

— 工作的性质和范围（合同要素的详情和它们的等级，关于与建设项目有关的改动或者延展的权利的内容）；
— 赔偿条款（出现与所描述的工作相背离的情况时的赔偿条款）；
— 执行条款（关于业主对工作的监督，以确保现场的正常秩序和施工单位对现场设备的正确使用的条款；施工单位对业主或者设计师的要求有意见的时候控制申诉权的条款）；
— 执行文件（关于移交与所执行的工作有关的文件的条款）；
— 时间安排（一般条款，例如，在没有明确日期的情况下，保证建设工作在规定时间内开始的条款）；
— 障碍（在近期内有缔结合法合同的障碍时确定程序的条款。例如，提前告知业主障碍的存在和它们的影响，从而采取应对措施）；
— 取消（关于业主或施工单位取消合同的条款）；
— 责任和义务（合同双方的责任）；
— 合同处罚（规定不够刑罚等级的合同处罚形式）；
— 承诺（工作时间安排的法律性承诺）；
— 担保（在建筑交付之后保证业主的要求的条款）；
— 协议（关于全部或部分工作完成之后采用何种方式、以什么顺序移交的规定）；
— 计时付费的工作（关于合同外的劳务报酬的条款，例如施工单位在这样的工作开始前与业主做的约定）；
— 报酬支付（关于分期付款、半最终发票和最终发票的一般条款，例如，时间表要一直延续到最终发票审核）；
— 保证金（关于合同双方的相互保障的条款，例如抵押物或者保证金）；
— 争议（关于出现争议的条款，例如确定业主的权限）。

特殊合同条款

特殊合同条款可以和一般合同条款讨论同一个问题，并且在某些地方对它进行补充。它们是一般合同条款的附属，而不是它们的替代。就比较典型的情况来说，如果一般合同条款已经有了原则性的内容，那么就会把特殊合同条款放在招标文件里。它们会涵盖下列内容：

— 发票（发票必须根据它们所收的款项分为分期发票、半最终发票和最终发票，而且号码必须是连续的。其他的格式要求可以根据工作进度处理，例如，根据对要求的工作的描述来确定）；
— 特殊支付形式（关于业主向施工单位支付报酬和与报酬相关的规定。例如，可以达成一个支付协议，确定支付的程度和日期。通常会在特定的时间支付到这个时间点为止的工作报酬）；
— 确定价格的根据（报价人用来确定价格的计算书）；
— 工资或材料灵活的价格条款（关于在施工过程中发生工资水平或建筑材料价格变化时，允许合同价格调整的条款）；
— 额外成本的通知（规定出现任何额外成本都必须及早通知业主的条款）；
— 分包商（分包商是提供公司本身不能完成的工作的。如果不允许或者只有在一定条件下才允许采用分包商，那么就应该在特殊合同条款中有所明示）；
— 竞争限制（不许在与提交或者未提交的报价、价格或者利润补充有关的报价人之间进行违规竞争的协议。特殊合同条款规定了采用违规竞争行为的后果）；
— 减价（通常会是一个百分比并且从所有发票中适当扣除）；
— 环保（通常不会规定具体的环保措施。习惯上特殊合同条款会参考减少环境破坏的建筑措施）；
— 合同变更（特殊合同条款要规定合同变更形式。例如，必须以书面形式对合同进行修改）。

P44 技术要求

通常，当达到了一定的详细程度的时候，建筑项目策划和对所需要的工作的描述就完成了。其他的所有东西都是根据技术要求来确定的。其中包括工作方式的说明。例如，设计师会提供一张钢筋混凝土墙的图纸，并且以详细描述框架、钢筋和混凝土处理的文字为补充。但是他们不会详细描述如何搭建框架、怎样安插钢筋或者怎样压实混凝土。这些内容是施工单位专业知识的一部分，并且由设计师通过招标文件中的技术要求传达给建筑公司。

技术要求可以看作是关于由不同行业提供的绝大多数服务的条款

提示：

如果某些规定适用于某个具体的建设项目，那么应该把它们写在与项目有关的合同条款里，而不是写在通常用来规定若干个建设项目的特殊合同条款里。

提示：

通用的技术规范是建立在一整套经过长期检验的技术规定之上的。更高的标准是根据最近的、但是没有经过测试和检验的技术水平来制定的。考虑最新科学成果的科技水平则可以提出更高的要求。

之和。>见“附录”它们包括以最低标准的形式对大量施工工作进行约束的相关规定。特殊技术要求则用来限定比较高的标准。

一般技术要求

一般技术要求是某种技术的通用标准。

这些规定通常是针对具体的行业制定的，包括适用范围、使用的物品和材料、施工方式、作为整体服务一个组成部分的附加服务等内容，以及财务和编写服务说明书的提示。

特殊技术合同条款

特殊技术要求可以是一般技术要求的补充，也可以是对先前没有规定的内容的要求。例如，特殊技术要求可以针对一般技术要求中没有提到的施工过程，也可以规定更加严格的误差要求以作为对现有最低标准的补充。

特殊技术要求是以规范以及其他技术规定、生产方针或者规定和相关部门的指令为基础的。

考虑到现有的技术状态和某项工作的科技水平，还可以提出更多的严格要求；这些要求都是以单个的许可证为基础的。

而且，还有中间状态的特殊技术要求：例如，当由于技术原因在施工过程中必须接受某些工作的时候，因为由于施工进度的关系在之后的阶段它们很难达到要求。

P46

与项目相关的合同条款

项目的相关信息涵盖了项目的概况。它们包括了影响建设项目的所有合同的和技术的规定。

关于建筑基地的内容

这些特殊合同条款必须针对每一个项目具体分析。它们应当包括下列与建筑基地相关的内容：

— 位置（地址和对建筑所处位置的描述）；
— 通路（如何到达基地）；
— 存储空间（供承包商在施工过程中使用的区域）；
— 抬升设备（起重机或者塔吊之类的抬升设备通常会布置在施

工现场，供不同的公司运输建材之用）；

— 脚手架（脚手架可以由其他公司来处理）；

— 接通电路和给排水（作为现场设备的一部分，应该在施工开始前确定适当的接口）；

— 卫生设备（如果有的话）；

— 废水处理；

— 接通电话。

一般施工现场成本分配

一般施工现场成本可以通过合同分配给所有相关的承包商。用来设置现场标志、使用现场设备和废水处理的成本也可以通过合同条款进行分配。

实施阶段/合同期限

关于允许的工作时间的规定是非常重要的。所有与之相关的规定都是根据项目来确定的。它们包括施工开始和结束的规定。这些阶段跟之后建设工作的实施密切相关，如果不仔细研究，它们就可能带来损失或者合同上的惩罚。只有在与项目相关的合同条款中提到施工阶段才是具有法律效力的约束。如果合同中只和施工单位约定了中间的时间期限而不是开始和结束的时间的话，就必须在相关的合同条款中单独确定每一个阶段的时间。

合同处罚

如果没有满足合同期限的要求，那么这通常就意味着业主会对施工单位提出索赔。在这里，只会考虑那些实际发生的损失。如果其他条款中也有关于违背合同的处理方式的条款，那么必须在与项目相关的合同条款中适当地提示参考合同处罚条款。

如果必要的话，也可以跟其他与合同相关的条款达成一致，例如关于由其他承包商提供的平行服务的规定，或者是清理施工现场的条款。

P47

招标说明书

招标说明书是招标文件的关键要素。功能性标书和详细标书之间的区别主要是根据招标说明书的特征来确定的。详细的标书要列出服务目录，而功能性标书则是列出服务的计划。除了特殊情况之外，一般都会把建筑描述用作功能性招标说明书。＞见图 23

最后一章会详细讨论编写详细的或者功能性招标说明书的程序。

P48

图纸部分

招标文件中所附的平面、图纸或者草图是为了让承包商能够更好地根据所提供的服务进行报价。因此它们需要所有的用来确定基本几

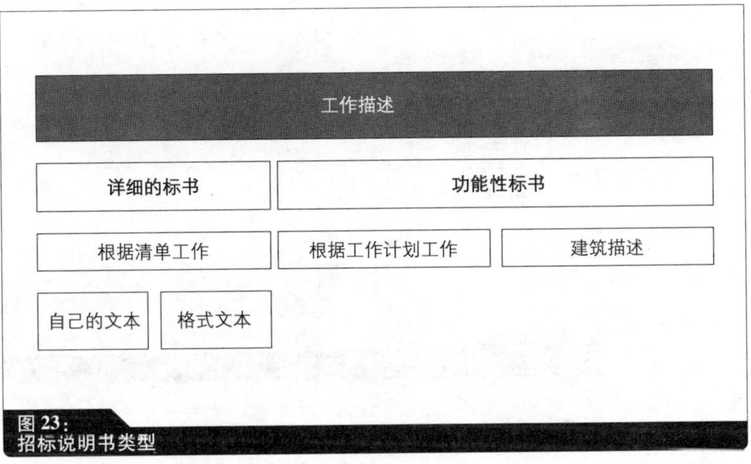

图 23：
招标说明书类型

何方位和理解任务要求的设计文件。

图纸部分的范围

用图纸描述的内容包可以是简单的手绘草图，也可以是技术图纸，平面的内容也是从总平面到按比例绘制的施工细部各不相同。> 见图 24、25

参考

建筑师可以在招标说明书中用某个需要特殊设计的细节的参考资料来说明描述文字中没有直接表达出来的东西或者是那些用图纸更容易说清楚的东西。

> **提示：**
>
> 建筑施工是以建筑师的施工图而不是招标说明书为根据的。招标说明书是编写投标报价的基础。如果招标说明书中的内容与设计相矛盾，那么应该对每一个部分进行仔细分析，按照适用的合同条款所确定的等级关系来执行。

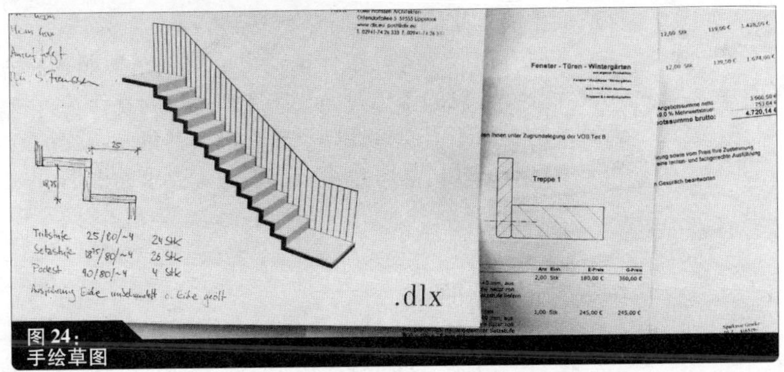

图 24：
手绘草图

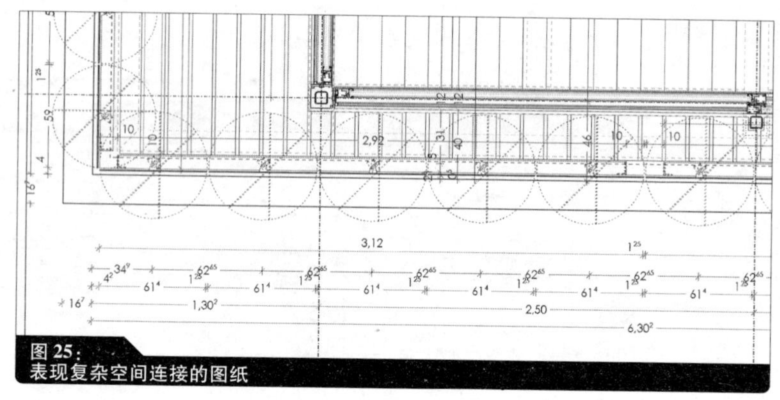

图 25：
表现复杂空间连接的图纸

P49

其他描述部分

样本 如果文字和图纸不能充分界定所需要做的工作，可以使用样本来说明。例如，如果钢筋混凝土需要一种一般的清水混凝土表面质量规定和要求中没有涵盖的特殊表面结构，那么做一个表面的样本并且把它们交给报价人，或者甚至直接把它放在招标文件中（比如说建筑中原有的木饰面板）会是一个比较明智的选择。

如果有样本，那么就需要在招标文件中予以充分说明，并且作出这样的总结："按照样本完成……"

制造工艺 招标文件中还需要提到制造工艺。在使用天然石材这样的情况下应该考虑制造工艺，因为石材的外观会有很大的差别。在这里，用文字的描述和图纸来说明几乎是不可能的。必须要求施工单位做一个材料的展示区，这样才能根据样品确定颜色、深浅以及内含物的特征和分布之类的重要标准。

也可以建立业主能够看到和感受到从表面材料到单个的家具和配件所产生的效果的展示室。

参照物，装修样品 在原有街区或者是整体中建设的时候，参照物特别重要。例如，在设计一栋建筑的外部时，砖的选择要与原有建筑一致，采用同样的砌筑方法，而不需要设计师解释规格、颜色或者胶粘剂。对原有建筑和那里已有品质的参考也可以作为施工要求的根据。

P51

编写服务说明书

我们把功能性的或者详细的服务说明书分成三种基本的初始

情况：

1. 没有提供设计，
2. 已经有规划或者规划许可，
3. 已经完成最后的设计阶段。

P51

编写功能性服务说明书的方法

目的

功能性服务说明书的目的是把所有对于一栋建筑来说必要的要求都放到一起。

编写功能性说明书的工具

编写功能性服务说明书可以在很大程度上依靠各种描述工具。它们包括：

— 建筑描述；
— 建筑任务书；
— 房间任务书；
— 所有工作的清单。

施工和装饰装修手册与服务说明书详细内容的关系更加密切。它们的语言基本上不是功能性的。它包括很多的具体要求，因此是与功能性招标文件的开放概念原则相背离的。但是，如果所确定的质量要求不需要进一步讨论，或者是在对某个房间进行装修或装饰时再来补充的话，这些手册可以用作功能性招标文件的一部分。> 见图 27

这里所列出来的工具既可以作为编写功能性说明书的基础，也可以作为说明书的一部分。没有设计，只能编写一份建筑或房间分配的任务书；如果想要编写列出各个房间的所有要求的册子（房间册）

	详细的服务说明书	功能性服务说明书	
策划阶段	有工作计划	有设计	没有设计
检查报价的办法	经济方法（报价格）	设计、功能、技术和经济方法	技术和经济方法

图 26：
服务说明书的特征

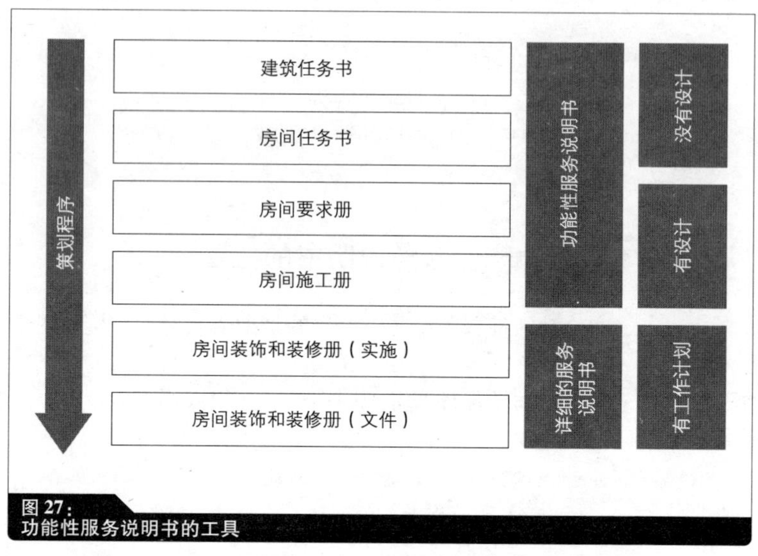

图 27：
功能性服务说明书的工具

就必须有设计。

不管是否有设计，都不会改变功能性服务说明书的基本原则。它的目的是要确定业主的所有要求，尽管如果没有提供设计的话，会要求报价人发挥更大的创造性。

P53

没有设计的功能性招标说明书

没有提供设计的建筑任务书或者房间分配任务书只是简单地描述了建筑整体、建筑的某个部分或者有特殊用途的区域的要求。由报价人负责设计、技术、用途和经济策划。

建筑任务书

建筑任务书要做出关于一栋建筑的基本说明。首先要详细介绍它的性质，比如说用途、办公室大小、办公类型、层数、每层的办公室数量或者有无地下室。

建筑任务书包括建设项目的概况以及其他与公共设施（污水、水、燃气、电和通信）、交通系统和外部通道的连接状况。

提示：

即使是有设计的时候，也可以把建筑任务书和房间分配任务书用作功能性招标说明书，只要它们没有彼此冲突。但是通常会从房间、结构构件和施工产品的角度对服务提出更加详细的要求。

建筑任务书必须制定使用面积的要求。这些要求可以进行局部改动和修正。例如，这个阶段甚至可以提出每个办公室的隔声要求。>见图28

房间分配任务书

房间分配任务书会提出更加明确的要求。它会给出房间和使用面积，以及如何布置和连接它们的信息。房间分配任务书中所包含的内容多少是根据策划阶段所达到的程度而决定的。根据下列标准进行合理的二次划分可以为整个描述打下良好的基础：

— 用途；

— 数量；

— 大小；

— 位置和朝向。>见图29

编号	范围	要求
I	区域	某某镇某某街12号，00001
II	要求（描述）	内城办公综合体
III	项目性质	改建
IV	用途	办公、小卖部
V	地块大小	10000m²
VI	层数	3
VII	地下室	有（1层）
VIII	建筑结构	两栋主楼，办公一栋辅助楼，小卖部
IX	办公空间	从____m²到____m²
X	小卖空间	从____m²到____m²
XI	办公类型	单间办公和开敞办公
XII	单间办公	从____m²到____m²
XIII	开敞办公	从____m²到____m²
XIV	通道	建筑与公共设施和交通相连
XV	停车	地下室可以停车，建筑北侧设停车场，每个工位一个车位
XVI	废物	集中废物处理
XVII	开放空间	有水池的花园
XVIII	条例和规范	开发计划，当地建筑要求

图28：
建筑任务书结构样本

功能配置

功能配置体现了各个区域彼此间的关系而不是说明它们所要求的面积。介绍各个区域之间的基本联系的目的是为了说明某种特殊用途的序列关系。> 见图 30

用途	数量	大小	位置和朝向
接待厅	1	$150m^2$	首层北侧/西侧
小卖部	1	$200m^2$	首层北侧/东侧
厨房	1	$80m^2$	首层中间/东侧
办公	4	每间$25m^2$	首层南侧/西侧
活动室	3	$1\times200m^2$ $2\times50m^2$	首层南侧/东侧

图 29：
图解式房间分配任务书实例

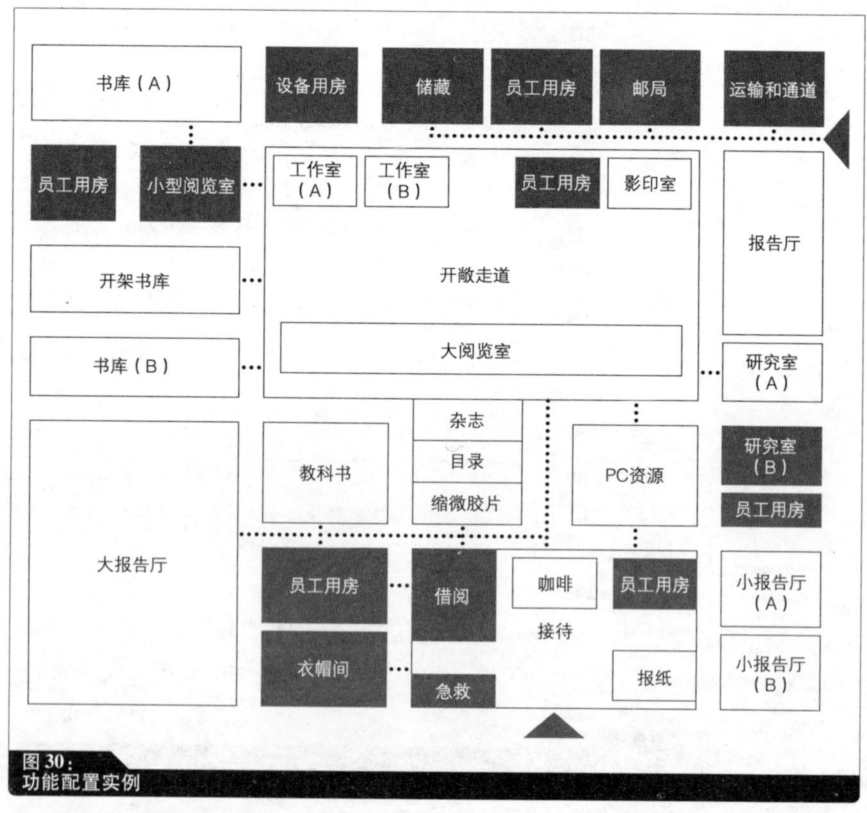

图 30：
功能配置实例

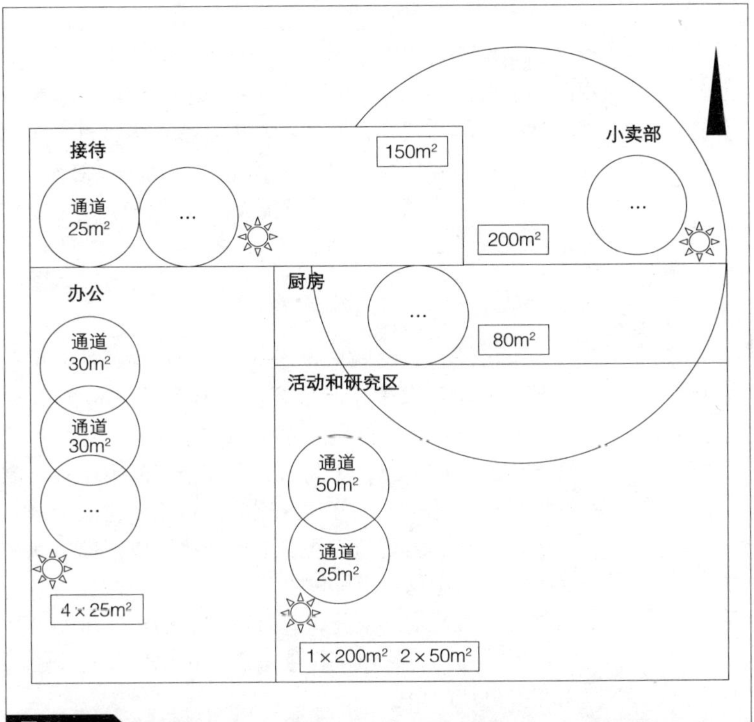

接待

通道
25m²

…

150m²

200m²

小卖部

…

办公

通道
30m²

通道
30m²

…

4×25m²

厨房

…

80m²

活动和研究区

通道
50m²

通道
25m²

1×200m² 2×50m²

图 31：
图表式房间任务书实例

图解式房间分配任务书 ✎

从表格式房间任务书和功能配置中得到的信息可以用一个图解式房间分配任务书来总结。它包括那些已经决定的基本建筑设计要素：根据所要求的区域和房间以及它们相互之间的连接关系提出正式的声明。> 见图 31

除了建筑和房间分配任务书之外还可以采用其他的描述方法。编写没有设计的功能性招标说明书时采用的建筑描述就是其中之一。

建筑描述

建筑描述主要提供整个建设项目的一个比较粗的概念，但是也可以传达功能性的细节。> 见“招标文件组织”，“文本部分”，“建设项目概况”等章节

小贴士：

可以用符号来增加更多的信息，比如说关于照度的数据（自然光/人工照明）来图解房间任务书。

跟以空间组织为基础的房间分配任务书不同，建筑描述是从施工或者行业的角度编写的。这也是它为什么比较粗略，只能作为功能性招标说明书的基础的原因。说“钢筋混凝土大厅，面积为2000m²”是一种非常概括的描述，但是对于功能性描述来说已经足够了。指定“钢筋混凝土”就排除了其他的结构构件，比如说钢梁。

建筑描述对某种结构形式确定得越细，就越能避免报价人选择其他结构，取得更高的效率。

P57

有设计的功能性说明书

如果已经有了设计，就已经有了建设项目的明确要求，它们已经在平面中体现出来了。但是要保证有设计的招标说明书的功能性特征，因为它仍然有可能确定单个结构系统、结构构件或者施工产品的质量。从这一点来说，房间册——某个房间中所有要完成的工作的一系列列表——是一个很好地把这些要求系统化的工具。

房间册

在策划阶段的适当时候引进房间册可以帮助房间分配任务书确定更多的细节。这是一个能够提供所有房间的信息的系统。房间册系统化地总结了所要求的空间，并且确定它们的使用要求。每个房间册都应该包括下列内容，以确保能清晰地识别，并且让这些信息的进一步使用变得更加简单：

— 房间数量，紧跟着建筑项目已经确定的系统；
— 房间定义；
— 空间特征；
— 要求或装修特点。

房间册类型

房间册的类型有三种，以满足不同目的和不同策划程度的要求：
— 房间要求册；
— 房间构造册；
— 房间装饰装修册。

提示：

房间册可以用作不同的目的：为设计确定策划要求；构成功能性招标说明书的基础；作为售楼文件；供施工阶段的现场管理团队所用；在完工时记录事情的状态以提供保障；或者是和建筑运行相关的用途。对于策划来说，它们还可以是很好的未来扩建或者改建措施的盘点工具。

房间要求册 房间要求册在有设计的招标说明书中起着至关重要的作用。每个房间或区域都有一张罗列了所有已知要求的表格。>见图 32

房间构造册 房间构造册提供了另一种表达方式，可以详细地描述某个房间的构造而不是装饰和装修的要求。>见图 33

房间册

房间要求册		页数：05	
建设项目：	Musterstadt South Weinreich Versicherungen（P45/145）		
建筑类型：	办公楼		
日期：	2007年5月3日	**准备人：**	穆勒先生
		批准人：	桑德斯女士
房间描述：	办公	**楼层：**	一层
房间号：	1.304		

技术要求		
区域	**要求**	**统计数值**
静力学 ……	最大误差 ……	t=1/300 ……
建筑科技 ……	防火每DIN4102 ……	结构构件至少为F30 ……

功能要求		
区域	**要求**	**统计数值**
……	……	……

设计要求		
区域	**要求**	**统计数值**
……	……	……

经济要求		
区域	**要求**	**统计数值**
……	……	……

生态要求		
区域	**要求**	**统计数值**
……	……	……

图 32：
房间要求册示例

房间册

		页数：	05
☐ 房间构造册		建设项目：	Oberstrasse 1 12345 Dorla
☒ 房间装饰装修册		建筑类型：	

日期：	2007年5月3日	准备人：	穆勒先生
		批准人：	桑德斯女士

房间描述：	办公	房间高度：	3.00m
房间号：	1.304	面积：	$20.60m^2$
楼层：	一层	类型：	使用

编号	元素	设备/结构	型号	数量
1	楼板	现浇钢筋混凝土板 冲击隔声层 找平层，地毯	C20/25 PE板 ZE 20, d=50mm	1
2	顶棚	现浇钢筋混凝土板 石膏板吊顶 水泥浆填缝 粉刷 ……	C20/25 …… …… …… ……	……
3	墙	……	……	……
4	窗户	……	……	……
5	门	……	……	……
6	照明	……	……	……
7	供电	……	……	……
8	采暖	……	……	……
9	通风	……	……	……

图 33：
房间装饰装修和构造册

房间装饰装修册

房间装饰装修册只是简单地提供每个房间装饰和装修的完整描述。它确定每个房间要安装哪些构件，以及它们的质量和数量要求。每个构件都按数量和生产商的描述，或者是比较详细的描述罗列出来。

P60

建筑细分目录

编写功能性招标说明书

为了按照功能把招标说明书系统化，应该对建筑进行细分，以提供通用的服务任务书——这和有完整服务列表的、针对行业进行划分

提示：

房间装饰装修册提供了每个房间的详细信息，主要是用于由每个行业完整的技术和一般要求列表的详细的招标说明书。应该仔细检查，以确保所有的服务都是按照行业、而不是按照房间列出来的。

的招标说明书正好相反。房间的记录清晰而系统，编号是连续的，按照它们在建筑和楼层中的位置排列。

为了进一步细分，可以编写一个使用或者功能分区列表。在这个层次上，它已经能够提供每个设备和技术要素、基础、承重结构、立面和屋顶的信息。如果已经知道或者计划好某个用途，那么甚至可以根据结构构件确定要求，比如两个办公室之间的非承重墙。> 见图 34

建设项目只需要细分到一定程度就可以了：必须能够根据业主的意图和已有的基本条件确定不同的要求。如果业主专门规定了一种功能，就没必要再把它细分成每一个构件了。

根据建筑或者房间分配任务书和根据房间册进行细分

上面提到的建筑任务书、房间分配任务书和房间册是用来提供细分目录的。它们可以通过要求的分配，根据不同的细分程度用作有服务任务书的功能性招标说明书。然而，并不是非要使用它们不可的。所有这些基本上都是建立确定要求的参考计划所必须的。它们可以是建筑的一部分，也可以是一个结构构件。

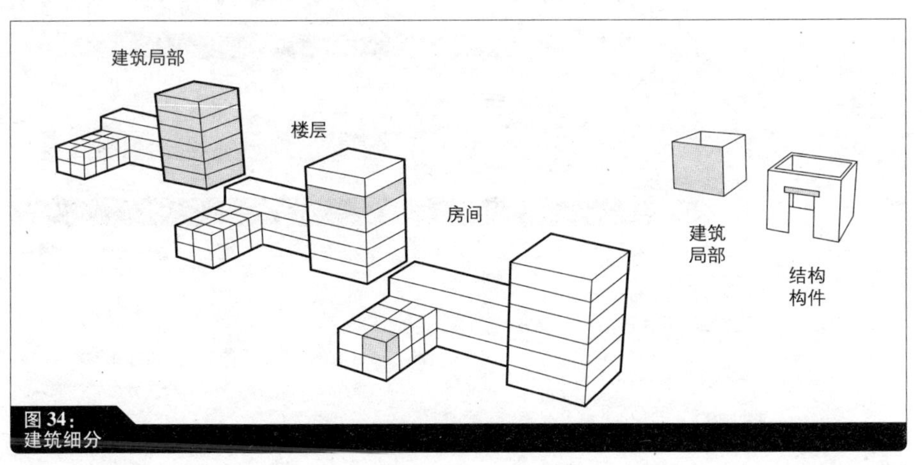

图 34：
建筑细分

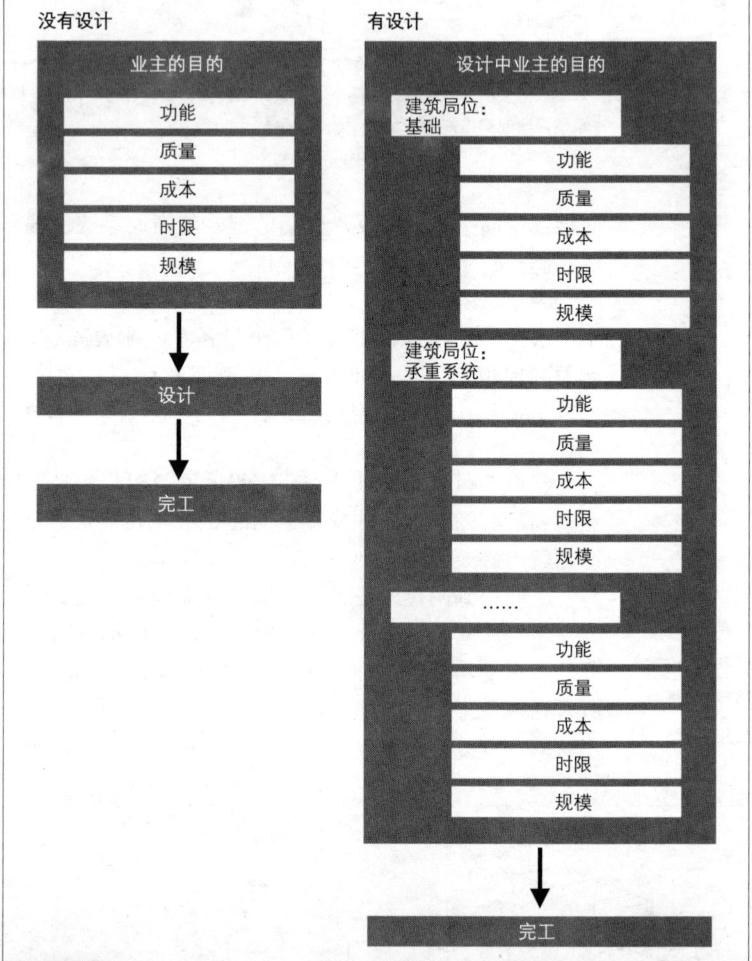

没有设计

业主的目的
功能
质量
成本
时限
规模

↓

设计

↓

完工

有设计

设计中业主的目的

建筑局位：
基础

- 功能
- 质量
- 成本
- 时限
- 规模

建筑局位：
承重系统

- 功能
- 质量
- 成本
- 时限
- 规模

……

- 功能
- 质量
- 成本
- 时限
- 规模

↓

完工

图 35：
参考计划

重点：

描述业主的意图需要进行深入的分析，这里面存在着很大的变数。也许业主只是想简单地确定某个产品厂家的生产量。那么报价人就要去调查一般情况下的所有不能改变的其他标准。如果这个过程是紧随其后的，那么既不可能也不应当打断项目的进程。

P62

草拟要求概述

建立细分系统之后，用途概况（甲方的目的和通用条件）就可以相应地运用到某一个最小的构件上（一个房间或者一个结构构件）。

建筑功能的粗略描述

我们建议首先把建设项目分配到一个功能组：

— 住宅；

— 办公楼；

— 商店；

— 学校，大学；

— 厂房；

— 医院。

要求分类和相关内容

用途概述可以通过设计（社会的和美学的）、技术、功能、经济和其他分类方式进行进一步区分。每个类别都有一致的要求。例如，建筑科学和构造都可以归入技术要求一类中。

单个要求和级别

也可以根据单个方面进行二次划分。这也是以要求部分的某个分项为基础的，可以更加精确地确立等级。单个要求通常包括参照标准和确定最小值的规定。有可能某一条建筑科学的要求是针对门的防火要求规定，它的最小耐火等级是 F30。

单个要求和隐藏在它们后面的等级可以用图 36 中的细分系统清楚地列举出来。

要求分类	要求方面	单个要求	数值
例如技术要求	例如建筑科学	例如防火	例如F60
			……
		……	……
			……
	例如静力学	……	……
			……
		……	……
			……
例如审美要求	……	……	……
……	……	……	……

图 36.
记录服务的系统实例

提示：

等级为每个要求建立了清晰的、重要的基础。如果没有给出等级，或者如果只是定性地划分了等级（比如说强调冲击隔声），也许会在报价人的陈述中出现很多不受欢迎的内容。

功能性招标说明书可能的表达方式

然而，在服务任务书中总结要求并不是惟一的表述方式。也可以连续地提出要求，只要能够保证具体的要求可以与某个特定的元素（例如一个建筑部分或者一个房间）相对应。

确定要求

要根据业主的希望和建设项目的基本情况充分、清楚地确定要求。应该按照上面所提到的分类和下面的参照，对可能的要求进行详细解释。

设计要求

设计要求既包括审美方面的要求也包括社会方面的要求。这一类型的内容在很大程度上体现了业主的感受，包括像从社会角度说的方便性、私密性和舒适性，以及从审美角度讲的建筑质量、优雅和影响力。

这些主观要求也许包括指定高质量的建筑材料或者是令人印象深刻的公共空间（例如一个宽敞的中庭）。

功能要求

功能要求也是根据业主的目的决定的。只不过是不同的功能组（比如说学校）会有不同的关键特征而已。这一类的要求会提到功能网格、屋顶跨度、楼层数、楼层面积、使用面积、首层平面的可变性或者是建筑的灵活使用等内容。某种技术特征的不同要求也是要遵循建筑的功能。

技术要求

技术要求直接来自于功能、标准和规定，或者是来自于业主表达的愿望。基本上，所有与承重能力、稳定性和建筑科学（例如采暖和隔声、防火和防水）有关的内容都可以通过要求确定得更加充分。

例如，技术要求可以遵循业主对空调办公空间的要求，必须在功能性说明书中予以考虑。

经济要求

与建筑相关的经济问题也在很大程度上取决于业主的目标。投资成本、建筑维护成本、运行成本或者使用率之类的问题都与业主的意愿以及所设计的建筑的功能和技术有着直接的联系。业主的战略目标之一可能是在于他或她的建筑有可选择的而不是固定的燃料，这也许意味着增加投资成本，但是可以在长期的运行过程中节约费用。

生态要求

所使用的建筑材料或者不破坏环境的能量概念的循环使用的可能

性、环境的稳固性之类的问题都包括在生态要求之中。这个分类的基本条件主要是一个合法性，但是也取决于业主的目的。业主也许会把国家对有利于环境的技术的补贴看作是采用这些技术的一个原因。然而，业主也有可能只是提到了要建一座低能耗住宅而没有进行详细地说明。其他从建筑的无污染转换或拆除角度出发的长远考虑也是生态要求的一部分。

功能性招标说明书的其他内容

除了上面已经提到的要求之外，服务任务书中没有包括的其他内容对于功能性招标说明书来说也是非常重要的。它包括法律和技术的规定、对建筑的粗略描述等从中能够发现当地对项目的要求以及明确在报价中不需要包括哪些工作的内容。

图 37 总结了可能的要求中最重要的关键点。

工作任务书必须把要求明确地提出来。可以用连续的文本、清单

图 37：
要求标准

或表格来表达。这在很大程度上取决于任务书细分工作的方法。如果任务书是从建筑整体一直讲到单个房间，就可以按照下列方式列出要求表：

建筑整体的要求：

— 两层；

— 实体结构；

— 与现行规定一致的低能耗住宅最低要求。

单个建筑局部的要求：

— 采用玻璃顶板的温室；

— 自然通风；

— 夏季室内最高温度：29℃。

单个房间的要求：

— 朝向花园的书房；

— 楼梯自然采光；

— 有单独厕所的浴室。

这种系统化的列表可以很快列出每个不同区域的工作任务。单个的要求可以给定一个指定的等级，但是设计要求很难精确地划分。可以很明确地规定酒店的每一个房间都能看到海景，但是像“舒适的气氛”或者“休闲风格”这样的描述就在很大程度上取决于个人的想法和经验，除非还有更加详细的说明，否则起不到什么作用。

P66

详细说明的工作程序

目的

可以用详细的工作说明来编写所要求工作的最经济报价。报价人必须根据已经完成的工作计划详细而全面地描述投标的途径。设计师会有一套测试系统。

结构

招标说明书是一个通用的结构系统，可以用它来连续地记录单项的工作。把单项工作在表格中逐项列出来，分批、分专业、分标题给出数量。在实践过程中，招标说明书的细分体现了各个位置和专业领域所需要做的工作。> 见图 38

根据位置和房间进行的二次划分是功能性招标说明书中非常常见的形式。另一种系统化的方式是根据行业进行二次划分。

招标说明书和合同单价

隐藏在招标说明书后面的原则是它要逐项直接询问最小描述元素的价格。报价人应该通过精确地描述特征、数量和质量来提供这些列项的单价。

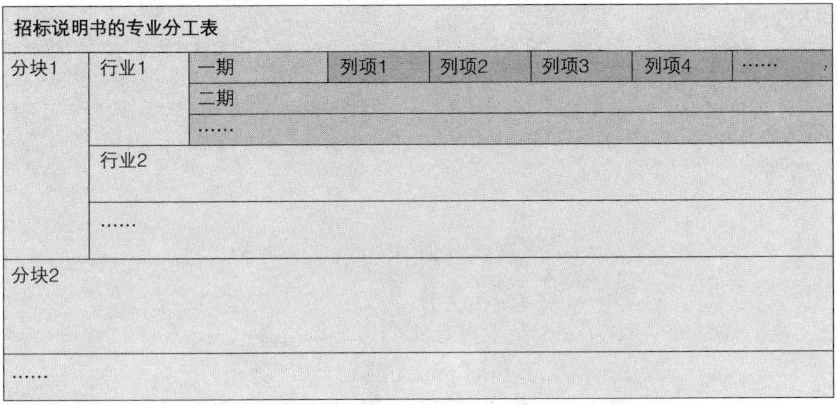

招标说明书的专业分工表							
分块1	行业1	一期	列项1	列项2	列项3	列项4	……
		二期					
		……					
	行业2						
	……						
分块2							
……							

图 38：
招标说明书的细分系统

把所需数量和单价相乘就可以得出每一项的总价。所有总价之和就是净报价。当之后要开具发票时，只引用单价，而不是计划的数量。这种类型的合同单价通常根据实际使用的数量来记账。

编写招标说明书

如果招标说明书需要一份对所需提供服务的描述，那么就必须把这些信息系统地从计划转变成根据行业排列的次作业，因为根据它们的特点来说，计划都是根据施工组成部分来编写的。

如果很清楚地列出了每项次作业，那么首先应该考虑工作的顺序和施工方法。采用这种方法可以很容易地把同样性质的次作业确定地更加具体，并且把它们分派给各个行业。可以通过回答下列问题来界定这些次作业、分派到每个行业、并且切实地对工作进行描述：

— 次作业必须描述的是哪个结构元素？

顶棚；墙；基础……

— 次作业必须描述的是哪种结构类型？

砌体；钢筋混凝土……

提示：

单价是一个单位为基础的价格，例如 10 欧元/m^2。

小贴士：

详细介绍装饰和装修的房间册是编写有工作列表的招标说明书非常有用的工具。它可以详细地给出房间的数量以及面积。因此可以在招标说明书的深入描述中快速而系统地记录某些工作的数量和质量。

提示：

在行业标准中能够找到建设工作的专业分工。这些标准中还包括了工作、建筑材料和所使用的结构构件、作为基础的单位、适当的次作业、发票开具和编写招标说明的分类信息（见“附录”）。

— *次作业必须描述的是什么样的程序？*

运土；钢筋混凝土工程……

— *建筑分期和不同行业之间怎么衔接？*

开挖 = 运土；基础 = 钢筋混凝土工程……

— *哪些次作业可以分配给特定的行业和施工单位？*

钢筋混凝土顶棚 = 支模、钢筋、混凝土……

下一步记录次作业更多的细节。在这儿我们建议把相关的规范、政策和规定总结一下，为详细地描述每一项次作业搭建一个框架，用建筑材料、结构构件可靠而专业的信息来源列出所要求的工作。

在这个基础上，可以逐步编写所有工作的招标说明书。

P69 分块

分块是分派给一家公司的完整的分包单位。地块可以看作是能够根据与空间（局部分块）或者专业服务（专业分块、行业）相关的标准确定的，平等而独立的分项目。

局部分块 根据区域分块的情况通常只出现在大型建设项目中，能够把一栋沿街建筑分成几个阶段或者施工合同部分。如果业主只想委托一部分建设工作而把后续工作分配给其他公司，那么他或者她就需要进行适当地分块。

专业分块 如果业主根据行业分工来分包合同，那么就会采用专业分块的形式。>见“招标文件组织”一章一个行业可以分割成几个专业分块。例如，可以委托金属制造工制作栏杆而由其他人来完成立面。

提示：

就像独立的建设项目一样，分块也可以从分区和与建筑局部或某个行业相关的专业领域的角度进行二次划分。也可以根据行业或标题确定分块，然后作为一个完整的工作包分包。

P69 标题和副标题

标题 标题是在整体项目的层次下面对建设项目进行进一步划分。标题描述了一个建筑部分或者分块或整个项目中的某个特定行业。它可以描述一个行业中的次作业，但是不在报价中提交自己完整单位的价格。和分块相反，标题的功能是在连续的章节中把单项工作总结起来。把属于同一专业或物理分区的单项工作（项目）打包，这样能够为在整体问题中确定单项的价格打下良好的基础。

例如，为了把工作分成连续的部分，“金属制造”招标说明书中也许会包括像“楼梯和扶手”、“门和门框”、还有“栅栏”这样的标题。然后可以用“室外楼梯和扶手”和“室内楼梯和扶手”之类的副标题进行更加深入的区分。可以根据单个的结构构件划分工作。“框架结构”的招标说明书可以分成“基础”、“楼板”、“砖砌外墙”、“室内砌体墙”、“顶棚”等标题，或者根据工作所在的位置进行更加详细的划分，比如说“厨房瓷砖”、“厕所瓷砖”，然后再分成“地砖”和“墙砖”。

细分的等级数取决于设计师。他们只要根据项目的复杂程度对标书进行划分就行了。用标题和副标题进行划分必须保证单元的连续性。它们除了能够让人更好地理解分配给每个单项的任务之外，还可以在比较报价的时候衡量总价里的其他东西而不仅仅是每个单项。而且，设计师可以根据标题比较报价。例如上面提到的金属制品的例子，它可以在比较报价的时候看出来某个金属制品商提供了楼梯和门最优惠的价格，但是围栏的价格要比平均价高出许多。

副标题 可以用副标题对单个标题进行细分。例如，某项特殊工作要求的钢筋混凝土可以总结在一个标题下面，而把它所需要的模板和加固材料列在副标题里。

标题和副标题或者其他细分等级（必要时候采用的主标题）之间到底有什么区别，以及工作范围和专业领域的再次划分采用什么样的顺序，这些都取决于单个建设项目的规模和复杂程度，以及与之相

小贴士：

分标题列举有利于评估单个部分或行业的报价。这样，可以在一个单独的标题下面列出所有项目的总价，同时提供每个标题的清单。

— 土木工程
— 钻孔
— 准备工作
— 夯土、滤网、压缩
— 水管
— 污水排放
— 排水
— 混凝土喷涂
— 道路施工
— 景观
— 喷涂
— 地下电缆
— 轨道施工
— 石材
— 凝固
— 天然石材
— 人造石材
— 木工
— 钢结构
— 密封
— 屋顶面层和屋顶密封
— 管道设备
— 干作业
— 复合保温系统
— 混凝土养护
— 抹灰和粉刷
— 呼吸幕墙
— 瓷砖和面板安装
— 找平
— 浇沥青
— 细木工
— 镶木地板
— 金属配件
— 百叶窗
— 金属制品
— 玻璃
— 油漆、抛光、饰面
— 钢铁防腐蚀
— 地板面层
— 壁纸
— 木地板
— 风道安装
— 采暖和中央热水设备
— 燃气、水和排水设施
— 中低频率设备
— 照明设备
— 传送系统、电梯、扶梯
— 建筑自动化
— 脚手架
— 拆除

图 39：
不同行业列表

关的合同的性质。图 39 体现了以各工种的专业分配为基础的划分可能性。

P71

列项

招标说明书中的一个列项是最小的招标单位，它代表了施工中的一个次作业。它由能够清楚而明确地确定要做的工作和特定的服务要求的描述性语言所组成。描述性的语言可以采用规定的格式或者按照标准分类放在一起。

可以在一个列项中包括一个以上的工作，只要它们具有相同的技术特征和价格计算方式。

标书列项的构成

在这个背景之下，我们建议在招标说明书中系统地描述所有的列项。从这个角度来说，标书列项的典型构成要素是按照下列类型分组的：

编号	文本	列项	数量	单位	单价	总价
01.02.02.0001	……楼板模板		50	m		

见图 40：
标书列项的构成

- – No. 编号
- – Text 描述性文字（或长或短的文本）
- – ITy 列项类型
- – Quantity 根据数量单位计算出来的数量
- – UQ 数量单位
- – UP 单价
- – TP 每个列项的总价（单价×数量）

这些分类可以应用于每一个列项，从而形成按地块、行业、标题或副标题排列的说明书。由报价人填写单价和总价。>见图 40

应该从招标的角度给出每个类型的下列信息：

编号

编号可以帮助我们查找招标说明书。根据工艺和价格归到同一类型中的每一项工作（次作业）都有一个按照既定的细分方式确定的专用号码。这个号码直接与项目的划分方式有关，并且在招标说明书中把它反映出来。在相对简单的项目中，可以按照下列方式进行编号。

地块	行业	标题	列项	索引
02	02	01	0001	a

索引

索引可以在编好号的列表中体现基本列项和可选项的关系。

编号在每个层次上应连续。>见图 41

招标说明的文本

招标说明书的文本应该按照招标文件的长短格式编写。

短文本

短文本主要用于编写报价和增加发票额的时候总结需要进一步提供的服务的。短文本必须要避免列项之间的混淆，必须清楚地确定每一个列项："石材 36. 5cm，地下室"、"石材 36. 5cm 首层"、"石材 17. 5cm 首层"等。

长文本

与之相反，长文本应该清楚地、竭尽所能地描述所要求的工作，这样所有的报价人都能正确地理解同一件事情。

自由格式和标准文本

从根本上来说，可以在考虑法律和技术规定的情况下自由组织文字。但是为了简化这个过程，设计师可以利用大家都能找到的标准化样本。

标准文本

标准文本是那些能够用来描述工作要求或者它的次种类的文本的集合。它们是根据预定的模式来编写的，包括建筑施工、建筑材料、

01.02.01.0001	1号地块行业2中标题1下面的1号活动
01.02.01.0002	1号地块行业2中标题1下面的2号活动
01.02.01.0003	1号地块行业2中标题1下面的3号活动
01.02.02.0001	1号地块行业2中**标题2**下面的1号活动
01.02.02.0002	1号地块行业2中**标题2**下面的2号活动
01.02.02.0003	1号地块行业2中**标题2**下面的3号活动
01.02.02.0004	1号地块行业2中**标题2**下面的4号活动
01.03.01.0001	1号地块**行业3**中标题1下面的1号活动
01.03.01.0002	1号地块**行业3**中标题1下面的2号活动
01.03.01.0003	1号地块**行业3**中标题1下面的3号活动
01.03.02.0001	1号地块行业3中**标题2**下面的1号活动
01.03.02.0002	1号地块行业3中**标题2**下面的2号活动
01.03.02.0003	1号地块行业3中**标题2**下面的3号活动

图 41：
使用编号的实例

尺寸和不同工作的数量单位等内容。

标准化说明文本最大的好处在于它们对所有报价人来说是一样的，在确定报价的时候不会出现不必要的工作或者额外的风险。而且，由于信息技术的兼容性，标准文本模式化的编写方式和等级化结构可以促进招标说明书的编写者和接受者之间的数据交换。设计师所要做的只是检查一下文本的正确性就行了。

可以在使用标准文本的既定系统的基础上最大限度地保证描述的完整性。但是需要注意的是，并不是所有的具体问题都有合适的模式文本的。

尤其是建筑产品生产商通常都会提供适用的格式文本，这些文本都可以很方便地应用在招标说明书中。设计师要清楚生产商这么做并

小贴士：

很多国家都采用标准文本来减少编写招标说明书的工作量。它们按照行业来排列，并且采用模式化的结构：对所要求的工作的描述性文字按照一个分成几个步骤的系统来编写，包括建筑类型、施工类型、施工质量、结构构件、材料类型、材料质量和尺寸以及施工条件等内容。但是无法保证文本在技术上的正确性，因为它有可能会变成没有什么意义的综合体。

无关紧要的独特卖点			
没有量化的描述	具体信息，例如负责检查的部门	针对产品和公司的描述	产品的相关特征
“……超过DIN要求” “……在平均水平之上” “……很好”	“取得……部门认证，证书号……”	“……人造橡胶沥青布线，上置PP织物……开VARIO槽，底部为薄膜。”	“采用挤压涂层工艺制造。

图 42：
生产商有偏见的描述

不是一种无私的奉献，而是把编写招标说明书作为产品推广的一种手段。他们还希望通过排除其他竞争对手的产品来建立自己独有的销售点。因此，可以通过提供包括与产品厚度相关的内容或者是指定每个厂家的特殊产品来达到在招标说明书中传达这些信息的目的，但是与产品的适用性和耐久性无关。这通常会为寻求更加合理的价格设置障碍，为招标公司带来过高的价格。

在任何情况下，设计师在采用生厂商的产品描述时都应该经过仔细考虑。如果他们无法确定某个信息在暗示什么，那么他们就应该咨询厂家或者更加中立的机构。

自由格式的文本

编写自由格式的文本需要有比较高的专业知识水平。这个方法也可以用来区分重要的和不重要的内容。应该咨询生产商、适当的协作单位或者其他的竞争对手，以获得与某个要求相关的信息。设计师必须根据相关标准和指令提出要求。这也适用于对整体的审核。为了确保这一点，描述文字必须涵盖下面的内容：

— 对所要求的工作的描述；
— 对所要求的工作的类型的描述；
— 可供参考的空间框架（关于建筑局部的信息，但是也关系到它在建筑中的位置，如果从编号中无法清楚地体现这些内容的话）；
— 关于质量的信息（材料、表面等）；
— 关于多大程度上偏离参考单元的信息。

系统化

下面的方案可以把对工作的文字描述系统化，可以用于工作中的所有部分。模式化的标准文本是以类似格式为基础的：

提示：

招标说明书可以建立在没有排他性的参照物的基础上（例如美国联邦技术规范局1076型不锈钢门把手或同等标准）。可以通过审核在报价中选用其他产品时提供的数据表或样品来保证是否符合同等标准。

— 建造方法、建筑类型（把建筑材料和构件放到一起后形成的产品）；
— 结构要素（构成一个房间或系统的建筑局部）；
— 建筑材料（想要的或者希望的建筑材料）；
— 尺寸1（构件尺寸，比如说墙厚）；
— 尺寸2（整体尺寸，比如说某项工作的安装高度）。

补充信息

也可以把关于发票或施工工艺的工作目标的内容安排在描述性文本中。例如，如果是钢筋混凝土建筑，那么采用预制构件会直接影响建造方式。同样，法律要求也可以作为招标说明书的一部分，例如完工时拆下来的材料归承包商所有。

参考

通常在使用文字来描述所要求的工作时会参考其他的文件，比如说静力学计算、平面图、样品或者报告。比如说，招标说明书的文本中会出现“按照加固计划加固”这样的语句。

举例：

建造方法：	按照 xxx 标准的砖石建筑
结构构件：	EG XX/YY 部分的内墙
建筑材料：	执行 yyy 标准的石灰质砂岩
构件尺寸：	17.5cm 厚
总体尺寸：	盖到 3.00m

可以通过补充与结构构件的质量（例如，两侧都采用清水砖墙）或者建筑材料（例如防盐水）或者建造方法（例如采用预制墙体）相关的信息来对这项工作进行深入解释。

提示：

如果不同的列项中出现了同样的描述，那么可以用两种方法解决不必要的重复。设计师可以在一个列项中全面地描述，然后在后面的列项中直接参考它（例如螺钉石膏板墙体，按前面的列项装修，但是采用双层板）。如果很多列项都采用同样的描述，设计师就可以把它们总结一下，写在前面的说明里（例如，螺钉石膏板墙采用基本说明中的A型装修）。基本说明总是和特定的工作联系在一起，广泛地应用于招标说明书的工作列表之中。基本上，它们和列举工作的文字一样，因此要在内容上和它们一致。保证一般合同条款和特殊合同条款不相互冲突。

在空间比较复杂的情况下或者比较复杂的建筑局部中，参考图纸是一个非常有用的办法。比如说，一个列项描述了有扶手的楼梯的构造，那么图纸可以帮助我们确定文本中提到的每一个构件并且清楚地理解它们的结合方式。这意味着报价人能够核对描述的完整性，并且更容易估计组装时间。

例如，在有特殊隔声要求的情况下，参考报告是非常有用的。在这种情况下，就不用把报告中已经确定的相关要求再单独列一遍了，而是通过适当的参照物来定义的："根据附带的隔声报告，新建筑的标准应高于建成建筑。必须满足这些要求，并在报价中予以考虑。"通过这种方法，设计师可以避免在把单项要求放到招标说明书中的时候出现错误。

列项类型（ITy）

招标说明书中可以采用不同类型的工作布置。设计师要把某一列项放到相应的"列项类型（ITy）栏"中。

不同列项之间的区别能够让设计师通过招标来检验可供选择的服务。例如，并不是所有在开始招标的时候所作的决定都是不能改变的（例如楼板面层的选择），也并不是每一项服务的要求都是一成不变的（例如排水管的安装）。可以通过随时增加的列项解决后续的变更。如果这些工作直到合同签署完之后才能确定，那么报价人提供的单价也是具有约束力的。

标准列项

标准列项都是已经实现了的。报价人可以在投标说明书中提供一个单价和总价。

编号	文本	列项	数量	单位	UP	TP
01.02.02.0001	织物地板面层	BI	30	m^2		
01.02.02.0002a	地毯	AI	30	m^2		

图 43：
可选项实例

可能出现的列项

可能出现的列项让设计师能够根据他们想要增加的服务进行市场检测。

这些列项无法确定是否会实施，投标说明书的总价是不包括这部分价格的，因此也不含在最终的报价内。如果业主在合同签署完毕之后要增加一个可能出现的列项，那么会根据报价人提供的单价来计算它的价格。如果没有委托他们做，那么可能出现的列项就不用谈报酬直接忽略了。

提示：

可能出现的列项只能用于不影响整个建设项目的辅助性工作中。如果某个列项在施工开始之前由于施工现场土壤条件资料不足而无法确定的话，采用可能出现的列项是一个明智的选择。

小贴士：

在分析报价的时候通常不会仔细检查可能出现的列项，因为它们不包括在总报价之内。这样就会导致在要实施可能出现的列项时被迫接受一个被夸大的价格。

基本的和可选的列项

基本列项是确定要实施的项目。也可以咨询可选项。因此，基本列项被看作是要提供的服务的一部分，必须提供单价和总价。

编号	文字	列项	数量	单位	UP	TP
01.02.02.0001	开挖土壤等级3-5		1000	m^2		
01.02.02.0001a	开挖土壤等级6	SI	200	m^2		

图 44：
补充项

如果所描述的服务是采用可选择的方式实施的话，就可以用可选的列项代替基本列项。因为对于可能出现的列项来说，可选项也必须有正确的界定，并且有报价者提供单价。>见图 43 策划师可以用这样的列项来作出最经济的决策。除非业主特意要求采用可选项，否则一般都会执行基本列项。

补充项

补充项是另一种不同的列项类型。它们确定潜在的障碍或者是根据标准列项需要增加的工作。报价人在投标说明书中提供补充项的单价和总价。相关的标准列项需要涵盖诸如列项的基本条件之类的内容，而补充项则描述更高标准或特殊安装条件。补充项的价格是根据较高标准和基本条件的差价来计算的。

提示：

可选项为承包商提供了一个提出自己的解决办法来补充基本列项的机会（例如，“根据之前的列项砌砖，但是根据报价人的选择执行”。）必须附加所提倡的做法的描述。

举例：

图 44 中的例子说明了隐藏在补充项背后的原则。在 01. 02. 02. 0001 中所说工作满足 $1000m^2$ 的 3 ~5 级土壤开挖。如果 $200m^2$ 的“六级开挖”有问题的话，就需要在适当补充项中进行询价。因此这个价格只包括解决问题的部分（补充价格），而不是代表六级开挖的价格。因此标准列项中已经包含了这个面积（$200m^2$）了。

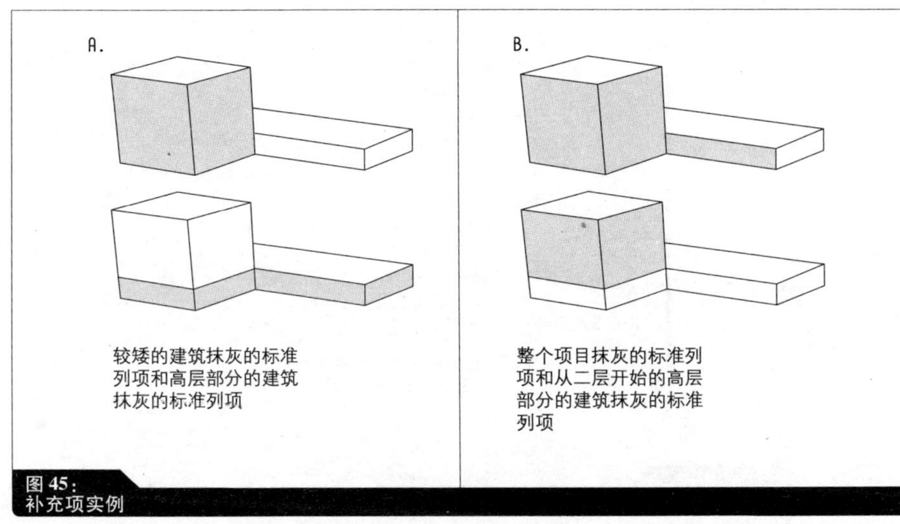

图 45：
补充项实例

可以用一个例子来说明采用补充项的可能性。图 45 中提出了描述同一项室外抹灰的两种形式。

第一种形式分别列出了高层建筑和低层建筑两个标准列项。另一种形式则在投标说明书中提供了整个项目的标准列项和较高区域的补充列项，其中需要包含脚手架的价格。第二种方式的优点是高层部分的首层不会因为额外的脚手架费用而增加报价。

数量和数量单位

数量可以用不同的单位表示。但是数量单位（QU）和某项工作的联系必须是有意义的。因此，也可以按照立方米（m^3）来计算钢筋；尽管从定形钢筋的角度来说是有缺陷的，工业计算标准才是决定价格的因素。合理的单位是与之相配的，比述说用平方米（m^2）来计算钢板，用米来计算钢条。

为某一列项所确定的单位是数量核查的基础，也是报价人确定某项工作的价格时的参考值。

小贴士：

通常补充项与相关的标准方式差别甚小。在这种情况下，描述性的文字可以简化成对主要改变的说明，而不需要重复一样的东西。只要写“xxx 列项补充说明：采用双层板”就足够了。

提示：

一般和特殊要求中会包括对数量单位的规定。例如，墙面抹灰按面积（m^2）计算，横梁按长度（m）计算。

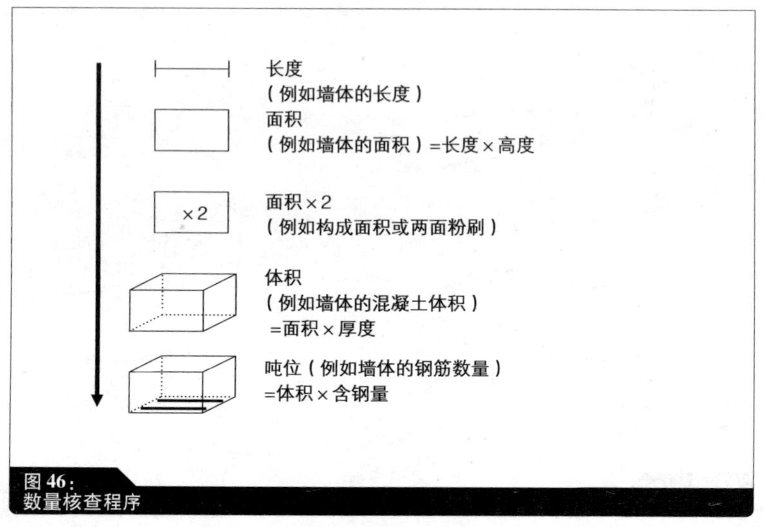

图 46：
数量核查程序

数量与按所要求的比例完成一项工作所需要的数量有关。数量是根据计划衡量完成工作所需要的数量的初步评估。它们直接来自于工作计划，在改建项目中也可以根据实地测量来决定，并且由设计师编写在招标说明书中。在一个特定的参考框架（整个建设项目、建筑局部、楼板等）中记录每个工作列项的数量，并且在计划中正确地用不同的颜色标记相关的结构构件以免造成不必要的重复，这是一个非常明智的办法。数量也可以用适当的结构软件来计算（例如 CAD）。从简单的面积到采用三维模型的复杂构件清单都可以通过自动数量计算而得到。

尽责的数量核查还可以简化后期的建设工作量的计算。

单价（UP）

单价是报价人根据工作的描述，按照单位数量计算并且编写进投标说明书的。通常单价是固定的，而且常常作为之后开具发票的根据。单价只有在数量或者工作内容发生巨大变化的时候才会改变。

总价（TP）

总价主要是根据产品单价和初步核定的数量（计划数量）来决定的。报价人必须计算出所有基本列项和预期会使用的补充列项的总价，并且把它放到投标说明书的适当位置。整个建设项目所有总价之和就是净的最终报价。加上法定增值税就是报价人承担各项业务的总价。总价中不含可选项和可能会发生的列项，因为在制定报价的时候无法确定这些项目是否会实施。

发票是根据所有单项总价之和实际发生的数量来确定的。

P82 结语

招标通常不是设计师所愿意承担的工作。这是可以理解的，因为一个完美的设计、精彩的观点甚至是认真策划的细节看上去无疑要有吸引力得多。大量的文字工作常常会破坏招标程序的吸引力。

但是为了实现设计想法而认真招标的设计师能够对自己的设计有更深刻的认识并且掌握实现它所必需的操作程序。

只有招标文件才能保证高标准的设计，这一点也反映在完美的实施效果上。

因此本书试图激励设计师制定自己的通俗易懂的招标文件，并且把它们编写得更有意义。如果承包商能够很好地理解招标文件，那么设计师就不会令自己失望。

P83 附录

P83 参考文献

Bert Bielefeld, Lars-Philip Rusch: *Building projects in China*, Birkhäuser Verlag, Basel 2006

Bert Bielefeld, Falk Würfele: *Building projects in the European Union*, Birkhäuser Verlag, Basel 2005

Institution of Civil Engineers, Association of Consulting Engineers and Civil Engineering Contractors Association: *Tendering for Civil Engineering Contracts*, Thomas Telford Ltd, 2000

CIRIA: *The Environmental Handbooks for Building and Civil Engineering: Vol 1. Design and Specification*, Thomas Telford Ltd, 1994

P83 国际合同样本

FIDIC	国际工程协会联合会
NEC	新工程合同

P83

其他资料来源

ISO	国际标准组织 （http：//www. iso. org）
CEN	欧洲标准委员会 （http：//www. cen. eu）

除了上面提到的渠道之外，还有许多提供传单、附加技术条款样本以及某些工作列项招标文本的样本的国家和国际组织。可以在下列网址找到各种承包商通用的标书样本：

互联网站

http：//publication. europe. eu
http：//www. neccontract. co. uk
http：//www. fidic. org

招标门户网站：

http：//ted. europe. eu

P84

图片来源

图 6 左	aboutpixel. de
图 6 右中	PixelQuelle. de
图 7	PixelQuelle. de
图 8	PixelQuelle. de
图 10 左中	aboutpixel. de
图 10 右	aboutpixel. de
其余插图	作者

尊敬的读者：

感谢您选购我社图书！建工版图书按图书销售分类在卖场上架，共设22个一级分类及43个二级分类，根据图书销售分类选购建筑类图书会节省您的大量时间。现将建工版图书销售分类及与我社联系方式介绍给您，欢迎随时与我们联系。

★建工版图书销售分类表（见下表）。

★欢迎登陆中国建筑工业出版社网站www.cabp.com.cn，本网站为您提供建工版图书信息查询，网上留言、购书服务，并邀请您加入网上读者俱乐部。

★中国建筑工业出版社总编室

电　话：010—58934845

传　真：010—68321361

★中国建筑工业出版社发行部

电　话：010—58933865

传　真：010—68325420

E-mail：hbw@cabp.com.cn

建工版图书销售分类表

一级分类名称（代码）	二级分类名称（代码）	一级分类名称（代码）	二级分类名称（代码）
建筑学（A）	建筑历史与理论（A10）	园林景观（G）	园林史与园林景观理论（G10）
	建筑设计（A20）		园林景观规划与设计（G20）
	建筑技术（A30）		环境艺术设计（G30）
	建筑表现·建筑制图（A40）		园林景观施工（G40）
	建筑艺术（A50）		园林植物与应用（G50）
建筑设备·建筑材料（F）	暖通空调（F10）	城乡建设·市政工程·环境工程（B）	城镇与乡（村）建设（B10）
	建筑给水排水（F20）		道路桥梁工程（B20）
	建筑电气与建筑智能化技术（F30）		市政给水排水工程（B30）
	建筑节能·建筑防火（F40）		市政供热、供燃气工程（B40）
	建筑材料（F50）		环境工程（B50）
城市规划·城市设计（P）	城市史与城市规划理论（P10）	建筑结构与岩土工程（S）	建筑结构（S10）
	城市规划与城市设计（P20）		岩土工程（S20）
室内设计·装饰装修（D）	室内设计与表现（D10）	建筑施工·设备安装技术（C）	施工技术（C10）
	家具与装饰（D20）		设备安装技术（C20）
	装修材料与施工（D30）		工程质量与安全（C30）
建筑工程经济与管理（M）	施工管理（M10）	房地产开发管理（E）	房地产开发与经营（E10）
	工程管理（M20）		物业管理（E20）
	工程监理（M30）	辞典·连续出版物（Z）	辞典（Z10）
	工程经济与造价（M40）		连续出版物（Z20）
艺术·设计（K）	艺术（K10）	旅游·其他（Q）	旅游（Q10）
	工业设计（K20）		其他（Q20）
	平面设计（K30）	土木建筑计算机应用系列（J）	
执业资格考试用书（R）		法律法规与标准规范单行本（T）	
高校教材（V）		法律法规与标准规范汇编/大全（U）	
高职高专教材（X）		培训教材（Y）	
中职中专教材（W）		电子出版物（H）	

注：建工版图书销售分类已标注于图书封底。